Praise for

Bliss Brain

'This book is a masterwork by a meditation adept who is also a scientist. Dawson captivates the reader with unforgettable stories and leading-edge neuroscience that open the mind, warm the heart, and elevate the spirit. His eloquent descriptions of meditative ecstasy and fact-based explanations of what brings it about provide big-picture understanding of the evolution of consciousness. This book shines with lucidity and caring. It most convincingly beckons readers to the heights of human awareness.'

~ **Judith Pennington**, president, Institute for the Awakened Mind

'This book is so superior to the dozens of popular books on meditation I've seen over the years that I want to shout about it from the rooftops! It will make a beautiful difference in people's lives! I opened it with the skepticism that its ambitious title invites, and I completed it surprised by how many fascinating insights I'd come to about the topic of "bliss brain".

At a minimum, readers who sense the potential benefits of meditation but feel they don't have a sustainable bridge into its practice will have lost that excuse. Dawson Church has brilliantly synthesized the vast literature on the neural, behavioral, lifestyle, and subjective benefits of meditation. He then systematically applies the principles derived from this analysis and from the fruits of his 50-year journey as a practicing and sometimes struggling meditator in formulating a disarmingly simple program.

But unlike most self-help programs, this one has been tested scientifically, and it reliably brings participants into states of consciousness and neurochemistry that previously required years of long, diligent practice. Bliss Brain *is a potent and urgently needed antidote to a culture in turmoil.'*

~ **David Feinstein, PhD**, co-author of *The Energies of Love*

'Science and love embrace in this incredible book. It's more than a clear and inspiring description of how radically our brains change in response to spiritual experiences; it shows us how to master the evolutionary quirks that sidetrack us in our quest for enlightenment. From one of today's most brilliant scientists and accomplished authors, Bliss Brain *is a gift to us all.'*

~ **Dianne Morrison**, president, Effective Action Consulting

'This brilliant book is a road map for turning tragedy into bliss. In my many years of knowing Dawson Church, I've found him to be a beacon of positivity and joy, a living example of the values he teaches. His research into how meditation activates the happiness centers of the brain puts Bliss Brain *within the reach of everyone. Dawson shows how you can grow resilience even in the face of great personal loss, using adversity to remodel your brain for bliss. This book will rock your world, and show you that happiness is truly achievable, no matter how far away it seems today.'*

~ **Lisa Garr**, host of *The Aware Show*

'Science has known for decades that the brain is changed by consciousness. Until now, we've never before had a truly practical guide showing how to put this knowledge to work. This comprehensive and accessible book integrates research from a dozen scientific fields and translates it for our everyday lives. It clearly shows us how to apply neural plasticity while giving us an evidence-based path to creating a happy, peaceful and resourceful brain. It's a "must read".'

~ **Marilyn Schlitz, PhD**, Dean of Transpersonal Psychology, Sofia University

'Stunning! Bliss Brain *is a masterpiece of science, inspiration and possibility; opening doors to each one of us about what's possible in our lives and unraveling the true meaning of human potential. I'll be reading this again and again and sharing it with those I love!'*

~ **Nick Ortner**, author of *The Tapping Solution*

'As The Secret *meets the scientist in Dawson's work, the boundaries of what you've believed possible will be stretched far beyond your existing picture of reality.'*

~ **Jack Canfield**, co-author of #1 *New York Times* bestselling Chicken Soup® series and featured teacher in *The Secret*

'The pace of change and the pressure to perform often leads professionals to the edge of burnout. The research in Bliss Brain *shows that meditation provides an island of inner peace in a stressful world. It leads to massive increases in productivity, creativity, happiness, and problem solving ability. The techniques Dawson offers us provide us with the leverage to address the challenges confronting organizations and society. These are the secrets that make companies and teams thrive, both personally and professionally.'*

~ **Dr. Ivan Misner**, founder of BNI and *New York Times* bestselling author

'Accumulating evidence suggests that meditation training yields improved emotional regulation and enduring changes in brain function, even outside meditation sessions, both in clinical and non-clinical populations. Dawson Church succinctly reviews all the neural evidence to date in this comprehensive yet relatable book, and shows how all of us can achieve states of bliss through meditation. Beautifully interwoven is his own story of loss and heartache, and recovery and resilience. This is a MUST read for everyone from all walks of life.'

~ **Associate Professor Peta Stapleton, PhD**, author of *The Science Behind Tapping*

'Four billion years of evolution have conditioned us to make safety & survival our brain's default Operating System. But now we're in the midst of an epoch change in which Enlightenment is emerging as the new Operating System. We're in the early stages of that transition and Bliss Brain *is the catalyst we need to speed up the process. In this book you'll find the precise instructions that science provides for reformatting your brain for lifetime happiness. Buy this book. You'll be oh-so-very-glad you did.'*

~ **Martin Rutte**, founder, Project Heaven on Earth

BLISS BRAIN

ALSO BY DAWSON CHURCH

Books

*Mind to Matter: The Astonishing Science of How Your Brain Creates Material Reality**

*The Genie in Your Genes: Epigenetic Medicine and the New Biology of Intention**

*Soul Medicine: Awakening Your Inner Blueprint for Abundant Health and Energy**

*The EFT (Emotional Freedom Techniques) Manual**

*EFT for Weight Loss**

*EFT for PTSD**

*EFT for Love Relationships**

*EFT for Fibromyalgia and Chronic Fatigue**

*EFT for Back Pain**

EFT for Golf

Psychological Trauma: Healing Its Roots in Brain, Body and Memory

The Clinical EFT Handbook Volume 1 (co-editor)*

The Clinical EFT Handbook Volume 2 (co-editor)*

*Communing with the Spirit of Your Unborn Child**

Facing Death, Finding Love

*Available from Hay House
Please visit:

Hay House UK: www.hayhouse.co.uk
Hay House USA: www.hayhouse.com®
Hay House Australia: www.hayhouse.com.au
Hay House India: www.hayhouse.co.in

BLISS BRAIN

The Neuroscience of Remodelling Your Brain for Resilience, Creativity and Joy

DAWSON CHURCH

HAY HOUSE
Carlsbad, California • New York City
London • Sydney • New Delhi

Published in the United Kingdom by:
Hay House UK Ltd, The Sixth Floor, Watson House,
54 Baker Street, London W1U 7BU
Tel: +44 (0)20 3927 7290; Fax: +44 (0)20 3927 7291; www.hayhouse.co.uk

Published in the United States of America by:
Hay House Inc., PO Box 5100, Carlsbad, CA 92018-5100
Tel: (1) 760 431 7695 or (800) 654 5126
Fax: (1) 760 431 6948 or (800) 650 5115; www.hayhouse.com

Published in Australia by:
Hay House Australia Ltd, 18/36 Ralph St, Alexandria NSW 2015
Tel: (61) 2 9669 4299; Fax: (61) 2 9669 4144; www.hayhouse.com.au

Published in India by:
Hay House Publishers India, Muskaan Complex, Plot No.3, B-2,
Vasant Kunj, New Delhi 110 070
Tel: (91) 11 4176 1620; Fax: (91) 11 4176 1630; www.hayhouse.co.in

Cover design: Victoria Valentine • *Interior design:* Nick C. Welch
Indexer: Joan Shapiro

A catalogue record for this book is available from the British Library.

Tradepaper ISBN: 978-1-78817-538-8
Hardback ISBN: 978-1-4019-5775-9
E-book ISBN: 978-1-4019-5776-6
Audiobook ISBN: 978-1-4019-5778-0

Printed and bound by CPI Group (UK) Ltd, Croydon CR0 4YY

This book is dedicated to my uncle, Alan Butler (1930–2011). His serenity and joy—continuously maintained while laboring under even the most difficult circumstances—showed me the way.

CONTENTS

FOREWORD

Scientists used to think that the brain you're born with is the brain you'll die with. While neuroscientists knew that muscles would grow with exercise, they believed that the brain is as good as it's going to get once our bodies are fully grown in adulthood.

Nothing could be further from the truth. Your brain grows and changes till you die. Every time you learn something new, read something new, taste something new . . . change happens inside your head every second.

Neuroplasticity is the law of the brain. *Neuroplasticity* means that what you do with your brain can actually produce changes in its structure. Whether you are aware of it or not, you are creating neural pathways in your brain by what you think, do, and feel.

In *Bliss Brain*, researcher Dawson Church digs into the latest studies on how what you do changes your brain. He also tells you exactly what to do to tune your brain to positivity. This ensures that the structural changes you produce through your emotions, thoughts, habits, and behavior are the ones that will support a long and happy life, instead of an early decline.

Dawson is hyper-focused on techniques that undo chronic stress and release emotions that keep you in a reactive state—responding to the past—rather than the present. He shows you how to achieve specific brain waves, the ones associated with relaxation and resilience. With these methods, you can reach a deep meditative state in minutes. He shows the remarkable effects this has on the body, from creating new stem cells to activating your mitochondria to lengthening your antiaging telomeres.

Dawson's meticulous research uses key biological markers to measure both stress and relaxation. He maps brain states with EEGs (electroencephalograms) and MRIs (magnetic resonance imaging), and uses indicators like cortisol and immunoglobulins to measure stress in the body. When he recommends a method, it's not based on guesswork or tradition; it's grounded in solid science.

Yet he also illustrates the science with stories. *Bliss Brain* is full of the stories of real-life people whose lives have been turned around by these techniques. War veterans. Incest survivors. Cancer patients. Trauma victims. People suffering from anxiety, depression, and other mental health conditions that limit their potential. Even a reformed drug dealer!

Personally and professionally, I'm all about high performance. I've worked a lot with EEGs, so I know its reliability and potential as a tool. With the right feedback, you can train your brain to achieve brain-wave states that enhance personal performance, inspire creativity, reset autonomic responses, and promote physical and mental health.

In my company 40 Years of Zen, we've even used EEG neurofeedback training with people who have suffered severe trauma. Veterans, car accident victims, and childhood abuse survivors can find relief through these methods. They can be trained to edit out the brain-wave patterns associated with trauma so that they're no longer being driven by it. Using the EEG readouts, we show them what their brain state is like when trauma is triggered, and also how to turn it off.

One of the reasons I started 40 Years of Zen was to increase my alpha brain waves. I don't think I could do the array of creative stuff I do if I didn't have the brain waves for it. And I absolutely know they are trainable. Dawson's techniques in this area intrigue me, and the fact that they can get me very quickly to where I want to be is a great gift.

In *Bliss Brain*, Dawson explains the range of brain-wave states, how each one changes the body, and how to turn on the ones that enhance brain function and overall health. The optimal brain state is one of "flow." EEG studies have identified it in spiritual masters, concert pianists, elite athletes, top creatives, and high performers.

Dawson calls this state "Bliss Brain" because of the flood of pleasurable neurochemicals that accompany flow. He describes the seven primary hormones and neurotransmitters produced by meditative flow, and how you can trigger their release by using specific mental and physical postures.

I'm a longtime meditator and I know the difficulties of trying to get into a calm mental space. Dawson explains why this is so hard, and why we can't turn off our brains when we meditate. Neuroscience tells us that brain activity doesn't fluctuate by more than 5% up or down, day or night, no matter what we are doing or not doing.

When you relax, your brain doesn't relax. Your brain activity stays at the same high level. When we're not doing a task, the brain sees all this unused spare capacity and grabs it.

These brain regions are called the Default Mode Network. And unfortunately, what your brain defaults to isn't happiness. It's ruminating about the disasters of your past and the possible threats in your future. That's why people have such difficulty when they try to meditate.

In *Bliss Brain*, we learn about EcoMeditation, Dawson's science-based method of getting to elevated states quickly. Research shows that these techniques quiet the Default Mode Network, then unlock that cascade of pleasure chemicals. I don't know of any other method that accomplishes this in such a short time—as little as 4 minutes.

Dawson reports on research showing that when people finish an EcoMeditation session, they maintain a flow state even after they open their eyes. This means that you can carry that elevated emotional state and all its benefits forward into your day—into your work, parenting, friendships, exercise, creative endeavors—into all the other parts of your life.

The tools in this book can dramatically improve your health and longevity. I've spent a ton of time, energy, and money on this subject, culminating most recently in my book *Super Human*. There are lots of things you can do to enhance longevity, but reducing stress, meditating regularly, and retraining your brain are keystones. They're central to Bliss Brain.

As I've interviewed and written about some of the top performers on the planet, people I call Game Changers, I've been struck by how many of them meditate. Most aren't professional meditation teachers; they're simply interested in optimizing their lives. It's the single most common practice described by Game Changers. In *Bliss Brain,* Dawson shows how you can make their best performance hack your own.

The trajectory of *Bliss Brain* is compelling. Dawson shares some of the personal disasters he's experienced, using his personal history as an example of the resilience that longtime meditation can bring. He shares how difficult he personally found meditation, despite joining a spiritual community as a teenager. In simple, clear language, laced with stories and analogies, he explains both the Default Mode Network and the brain's Enlightenment Circuit. He shows how science charts a path to attaining deep flow states quickly.

Dawson illustrates how the software of mind shapes the hardware of brain within 8 weeks of starting an effective meditation practice. How it produces those delicious bliss molecules that flood our brains as we enter flow states. Dawson describes each region of the brain changed by meditation, and concludes with a vision of human flourishing that will inspire and uplift you.

Make Dawson's "happiness habit" yours, use the tools in this book, and it can transform your life. *Bliss Brain* shows you the path that releases the natural potential of your brain for bliss.

— Dave Asprey, author of *Super Human: The Bulletproof Plan to Age Backward and Maybe Even Live Forever*

CHAPTER 1

FIRE

"Something's really wrong!"

My wife, Christine, is shaking my shoulder as she calls me out of a deep sleep.

I look groggily out of our bedroom window as she points at the midnight sky.

There's an orange glow on the horizon, starkly outlining the ridge opposite our house. The alarm clock winks 12:45 A.M.

I stumble out of bed, open the sliding glass door to the patio, and walk outside. A massive fire crests the opposite ridge and begins racing down the valley toward us.

I yell to Christine, "We're getting out of here RIGHT NOW."

I grab a T-shirt, pants, and a down jacket.

The electricity dies, and all the lights go out.

Shoes or socks? No time for both . . . shoes.

I run to the kitchen and fumble for the car keys in the dark. "We're taking the Honda!" I shout to Christine. Making a quick detour, I sprint through the living room to grab my laptop. Christine and I run from the house.

Close the door? Lock it? No time. Every second counts.

The cars are parked near our office building at the back of the property. Clouds of burning white embers swirl across the driveway, like flakes in a surreal snowstorm.

"Am I overreacting?" I wonder.

A huge gout of flame flares up right behind the office. It's shaped like a candle flame but 10 yards high.

No, I'm not overreacting.

ESCAPE

We leap into the Honda and I rev the powerful 271-horsepower engine to the max as I tear down our long driveway faster than I've ever driven it before.

I grip the steering wheel with white knuckles. The headlights are on, but I can't see the black asphalt because of the thick white smoke glowing in the headlights. I worry that I'll lose control of the car on our own driveway because of the insane speed I'm driving.

I slam on the brakes as we reach the road. There are other cars fleeing west on Mark West Springs Road, the haven where we have lived for a decade. Wait my turn? Be polite? That's me.

No time for polite. No time for the usual me. I barge into the line, forcing our red Honda Crosstour between two other cars.

Christine feels heat on her head, and is puzzled. She looks up through the moonroof.

All the tree branches above her are on fire.

Two miles down the road, I know we're out of immediate danger. What to do next? My ex-wife lives just three miles away, but the fire is traveling in her direction. Christine and I decide to drive to her house to make sure she and her family are alerted to the danger.

I ring her doorbell, but no one responds. I duck through the back gate and find a side door unlocked. I walk through the house, but it's empty and dark. She travels frequently, so she must be away.

Back outside, I wonder about waking the neighbors. We're three miles from the fire and it might never get to them. And I'm a polite person . . .

Not tonight. I lean on my horn.

A sleepy neighbor appears, bleary eyes blinking in the glare of the streetlight. Gray hair, Coke-bottle glasses, striped gray-and-black pajamas. I tell him we just fled a fire. He quickly wakes other neighbors who haven't yet been roused by my blaring horn. Within minutes, people are gathering prized possessions, documents, and pets, and packing their cars.

As Christine and I stand next to the Honda and look back in the direction we came from, the fire crests a hill a mile away. It's moving slowly now. With no electricity, it's the only light in that neighborhood.

We watch as it silhouettes a huge, $3-million mansion on the hillside. It burns all the vegetation around the house, but then moves on, leaving a blackened circle around the undamaged house. A fiery yellow ring around a black circle, with the house dead center.

Suddenly the left roof eave catches fire. Then the right eave. The house explodes in an enormous round bubble of flame.

It sounds as though we are in the middle of a battlefield. Car gas tanks explode as the fire touches them. Household propane tanks incinerate with loud booms. The hills magnify and echo the crashing sounds.

The fire is still moving slowly west toward us. We decide to drive to the home of our friends Bill and Jane, who live in Forestville, 20 minutes farther west. We get back in the Honda, its red skin pitted with white splatters where the embers hit.

Now the roads are gridlocked. Whole neighborhoods are fleeing. It takes us half an hour to reach the 101 freeway, a journey that would normally take 3 minutes. There, a police officer tells us we can only drive north, even though we need to go south on the first leg of our brief trip to Forestville. We know the fire is coming from the northeast, and he is probably sending people in the wrong direction, but we are powerless to argue. He seems more scared than we are.

We're locals, so we know how to evade the misleading blockades of the main roads and take back routes. We arrive at Bill and Jane's home. We feel awkward about waking them because it is now 3 A.M. We sit in the red Honda for a while, wondering what to do. Eventually, I ring their doorbell, but no one responds. Then our sleepy-eyed friends open the door. We tell them what happened and they are instantly wide awake.

After a debriefing, we go upstairs to sleep in their spare bedroom. We can't fall asleep, though, and go downstairs again. Jane has turned on the TV, but the only news available is from San Francisco, far from the fire area. Bill goes to the garage and sits in his car, listening to local radio. We phone Christine's daughter Julia who lives nearby in Petaluma, and she gleans fragments of information online.

Frustratingly, there is no hard information about how big the fire is, which direction it is traveling, or what residents of Sonoma County should do next. For a long time, Christine and I talk to Bill and Jane, pooling our ignorance.

TWO HUNDRED DOLLARS

We need some actionable information about where the fire is headed so that we can avoid its path. The news has only fragmented reports from isolated correspondents describing one horrific scene or another at some particular location. We know the fire is heading west, toward Forestville, toward us. It's 4 A.M.

The phone rings. It's an automated evacuation advisory for Forestville. Bill starts making phone calls, looking for a place for us to go. All the hotels are already full.

Eventually, he finds two rooms at Fort Ross Lodge, a hotel all the way west at the ocean. It's the farthest away from the fire we can get without swimming across the Pacific. Late that morning, we begin our preparations to drive to the coast.

Bill gives me a Trader Joe's bag filled with old clothes of his. Tie-dyed T-shirts and hoodies. Not my style—but I'm no longer a card-carrying member of the Fashion Police.

Jane opens her closet and tells Christine to take whatever she needs. We borrow suitcases from them so we have some place to put our very few material possessions.

We walk outside. Tiny flakes of ash fall all around us, like a gentle snow. These ash flakes are all that remains of schools, homes, shops, trees, gardens, and dreams.

Bill and Jane get in their car, we get in ours, and start driving.

The local market is still open, so Christine and I stop while Bill and Jane go ahead. The cashier tells us that because Forestville is being evacuated, the market is closing shortly. They aren't taking credit cards, only cash.

We try and force our swirling minds to think rationally. How much do we have? We count. Between us we have about $200. Will this $200 have to last us for a month? How can we store refrigerated food? What won't be available at the coast? What to buy right now?

Some items are obvious. Toothbrushes and toothpaste. Combs. Soap.

But people have been descending on the market as they flee and many shelves are empty. What was in those empty slots? What did we miss? We panic. We wonder, "Might this be the last food available for days?" We've been listening to the car radio, but there is still no solid information about what's happening.

I buy six pounds of cooked chicken sausage, a box of nutrition bars, and some fruit. Christine has no glasses. She can't remember what strength readers she uses. She buys three, each of different strengths. But what if those are wrong? She adds a different three to her collection. With our crazy collection of sausages and reading glasses, we drive to the coast.

On the radio, the announcer says that three people have died in the fire. My heart sinks: I know in my gut that the number is far too low. We were among the last people to flee Mark West Springs. There were no sirens, and the county's disaster warning system had not been activated.

THE DEATH TOLL

Later we learned that the county official in charge had decided that if the alarm was raised and everyone received mobile alerts, panic would ensue and the roads would get clogged. Because of that fateful decision, people like us in the path of the fire were never warned.

The final death toll was 22. Eight people died within a thousand yards of our home. Some died in their beds. Some died in their garages as they frantically tried to start their cars. Their fate was decided in seconds.

We spent the rest of the day with our friends, still hooked on the news, of which there was little. Emergency services and the broadcast system seemed to be in equal chaos. I texted Heather, the genius who runs our organization and who lives near Mark West. She'd been in touch with our team members and found that they were all safe, though they'd been evacuated.

Heather helped me compose a message for my blog on the Huffington Post, letting the international healing community know that we'd survived. For the next few weeks, I posted updates regularly.

In the afternoon, the sky grew dark long before sunset. Ash blowing from the fires covered the area like a gray mist. The sun glowed red and huge, like a tropical sunset. Toward nightfall, it turned crimson as it dipped below the horizon, and the first few stars twinkled through the gloom. That night, we sat in Bill and Jane's hotel room having a makeshift meal and drinking wine while watching the news. Every so often, Christine began to shake with tears. We lay together on the bed and I held her close.

We met some of our fellow fire refugees at the hotel, including some four-legged ones. The hotel had relaxed its "no pets" policy. One lady was taking two unwilling cats for a walk on a leash. They didn't like it. Talk about "herding cats."

The next morning, Christine and I had brunch with a group of people from the resort. We learned that overnight, the fire had jumped the 101 highway and destroyed the western neighborhood of Coffey Park before being contained by firefighters. It seemed very unlikely that our house had survived unscathed.

Leaving the Cats

10/17/2017 03:58 P.M. ET - Huffington Post - Santa Rosa Fire Blog Post 4

In a desperate attempt to get to our Honda and drive away from the approaching firestorm, we ran past the garage. That's where our two white Siamese cats, Pierre and Apple, spend each night. They're twins and we've had them since they were furry little kittens.

While I was running, my mind was working furiously: Is there time to get the cats?

A 10-yard-high fireball erupted behind the office building. There was no time to do anything but jump in the car and escape.

We slowed down after driving two miles. We hadn't said anything, but we were thinking the same thing. The cats.

1.1. Apple and Pierre cuddling.

Christine said, "Maybe the fire will go around the house. It does that sometimes. Maybe they'll find a way to get out of the garage. Wild animals are smart in a fire."

"That could happen," I reassured her, knowing in my gut that nothing would survive the inferno we had just driven out of.

When we saw the photographs of the ashes of our home the following day, we knew that the cats could not have survived. A *Wall Street Journal* report said that the fire was moving the length of a football field every 3 seconds, so the fire must have engulfed them quickly. Car gas tanks were erupting all around them; one of my classic cars was thrown 20 feet by the explosion. The most comforting thought I had was that the cats had died quickly.

In the few days since the fire, we've shed a hundred times more tears for the cats than we have for the loss of all our possessions and our home.

I've replayed the scene many times in my mind. Running to the car. Running past the garage. Wondering if I could have saved them.

Each time I run the mental movie, I wonder if I could have done something differently that awful night that would not have resulted in their deaths. I know the answer is no, but I keep looking for a crack in reality that will let me rewrite the past.

I know logically that if I'd diverted my attention from getting Christine and I out of the fire and blundered around in the dark garage trying to corral frantic cats, I would not be around to tell the tale. Neither Christine nor I would have survived. That analysis doesn't help me feel any better.

We'd taken precautions against wildfires, following the advice of local officials to trim back vegetation 100 feet around the house. But the morning before the fire, I had been washing my cars and looked at the vegetation around me. I thought, "This grass is more parched than I've ever seen it in the last 10 years."

The first day was spent in a flurry of communication, letting friends and family know that we were not among the people who had perished in the fire.

Though the National Guard was closing roads around the burned perimeter, allowing no one in, Heather managed to drive past our property. "How bad is the damage?" I texted.

"There's nothing left," she texted back. She sent us images of the destruction. Only the chimney remained, standing like a forlorn sentinel among the ashes. Even metal landmarks like filing cabinets and kitchen appliances had melted in the furnace. Heather and Ray's nearby house had survived. It was one of only six in her neighborhood to make it through the fire.

1.2. After the fire. See full-color version in center of book.

1.3. The same perspective 2 years before the fire.
See full-color version in center of book.

THE COLOR OF ASHES

Before the fire one of my most fun hobbies had been collecting classic cars. I had whittled the collection down to just a few of which I was particularly fond. I had two 1974 Jensen-Healeys, one red and one white. This classically beautiful British sports car was designed in the late 1960s and hand built in West Bromwich, England. I also kept an Italian 1980 Fiat Spider, my all-time favorite for driving the windy roads in wine country. Finally, there was a magnificent Rolls-Royce Silver Spirit, still looking factory new after 40 years, a tribute to the vision of the builders of "the world's finest motor car" and an example of why half the Rolls-Royces built in the last 100 years are still on the road.

Now only burned-out hulks remained.

The color of the photographs of our site was a uniform gray-brown. They looked as though they had been artistically sepia-toned. The intense heat had turned everything to the color of ash. The National Guard announced that it would be several weeks before residents would be allowed to return to their properties.

1.4. Remains of cars.

Our children were desperate to see us. Even though we had talked to them on the phone, it wasn't enough. They needed to be hugged and held and tousled and handheld and then hugged again. To celebrate the reality that we were alive. Christine's daughter Julia and her husband, Tyler, moved out of their small Petaluma apartment into a friend's apartment so that we could stay in theirs while we figured out what to do next.

THE MISSIONARY BARREL

The first order of business was clothing. Christine and I were dazed, like shell-shocked refugees fleeing a war zone. Julia and Tyler treated us indulgently, as though we were young children. They walked with us to a nearby high school gymnasium that had been hastily converted into a clothing repository and shelter. My son, Lionel, flew cross-country from New York while my daughter, Rexana, drove up from her apartment in Berkeley, California.

Christine's other daughter, Jessie, along with the other "kids," formed an impromptu rescue committee. Hours before we arrived from Fort Ross, they organized long To-Do lists using apps on their phones, with each one committed to handling some piece of the puzzle.

Find out where the local shelter is. See if clothes are available. Contact the insurance company. Cancel the phone and trash services at the property. Buy chargers for our laptops and cell phones. Find out how to get emergency passports, so we could travel to Canada where I was due to keynote the annual Energy Psychology conference that coming weekend. Buy toiletries and underwear. Find suitcases and storage containers. Locate soup kitchens.

That night we went out to eat at the local pub, which was serving free food to fire survivors. I hugged all the kids and we cried together and laughed together. We celebrated being together. I said, "I want us to be like this with each other every day, not just when there's an emergency."

1.5. The circle of love.

The staff at the shelter were kind, pointing out items I might need. I kept refusing, thinking thoughts like, "I don't need a shaver, I'll just grab my spare one from the RV." Then I'd realize that I'd seen our recreational vehicle in one of the photos. All that remained was the skeleton.

Sifting through the clothes brought back painful memories of my childhood. My parents moved to Colorado Springs when I was 4 years old, after serving as missionaries in Africa for many years. The church gave them a tiny cabin to live in, but they had no money to spare. The way my sister, Jenny, and I were clothed was from the "missionary barrel" at the church. There, parents left their children's cast-off clothing for those less fortunate.

I had bitter memories of the missionary barrel. My mother wouldn't allow me to take too much from it, and those old instructions, reawakened by the trauma of the fire, backwashed out of my subconscious mind.

I remembered a time when I found a stylish, warm boy's jacket in the missionary barrel. Beige with royal-blue horizontal stripes on the arms. It was just the thing for the Colorado winters, and I proudly claimed it as my own.

My mother came down hard on me. I had committed the sin of pride. Pride is a mortal sin. Mortal sins are the kind that get you damned to hell for all eternity.

Fortunately, my mother pulled me from the brink of the lake of fire by insisting that I put the jacket back in the missionary barrel the following Sunday. It was replaced by a shabby, worn, pilled, scratchy orange polyester parka four sizes too big for me, supplemented with neon-green pants.

In this weird assortment of mismatched, ill-fitting clothes, I made my first foray to kindergarten. I had a funny accent, dirt-poor parents, strange food in my lunch pail, and bizarre clothes. Not a good combination for the first day of school. I learned how merciless children could be.

The teachers decreed that my British accent must be fixed, so I was sent to remedial speech classes. Under the harsh attempts at correction, I developed a speech impediment—a severe stutter—and a social phobia. One of my early memories of Colorado Springs is walking through the snow to school, staring down at my feet as they carved trails in the slush. My heart lay heavy in my rain boots, as I foresaw the ridicule and isolation I would face the coming day.

With the Colorado Springs memories flooding back, I wanted to cry. But my beloved Christine was weeping freely, so shaky that the children had to hold her up. She needed a rock to stand on, so I kept my own grief in check.

The shelter had mountains of donated clothing, but none of it fit me because I'm so tall. The shelter director said, "We have a volunteer who's your height. I'm going to phone him and see if he has any clothes that might fit you." The next day I walked away with a huge duffel bag of clothing over my shoulder, my heart overflowing from the hugs and kindness of the people there.

1.6. At the evacuation center.

Another man, hearing that I was a public speaker and now had no jackets or suits, handed us $1,000 in rolled-up $20 bills and told me to go shopping.

Bless his heart. If I was going to go to the conference, I wanted to look great and not carry the energy of loss or misfortune. I used the $1,000 to buy two new outfits and my confidence swelled to fill the suits.

THE WATERSHED DECISIONS OF OUR LIVES

Christine and I couldn't go back to Mark West, and we couldn't do much else other than spend time with loved ones and answer emails. I was due to leave for Canada the next day. It would still be possible to make the trip if I left a few days later. But that would mean leaving Christine, who was still in a state of shock, as well as all the generous friends and family members who were helping us.

1.7. Me and the other giant.

I reached deep inside for the answers. Should I go or should I stay? It felt like one of the most critical decisions I would make in my entire life. A watershed. After the keynote in Vancouver, I was scheduled to teach a week-long training to therapists.

The topic? Psychological trauma.

My expertise in that field had just expanded exponentially.

From Vancouver, I was due to travel to New York and a couple other cities, lecturing and training health professionals. No one would complain if I canceled the trip, given the circumstances. Yet there was also very little I could do by remaining; our wonderful committee of twentysomething children were far more adept at locating resources and implementing plans than was I.

I prayed and sought inner guidance.

DOING YOUR LIFE'S MISSION

I decided to go.

After all, that's my life's mission. That's what I do. I realized in that critical moment how profound my commitment to that mission is, because I decided that even the loss of my office and home was not going to deter me from the work I was born to do. I still wonder if I made the right decision, but once the decision was made, the wheels went into motion.

I needed a passport and I needed it in 72 hours. Christine and I drove to the passport office in San Francisco, and the next day we had our new passports. People were reading my first blog about the fire and posting comments saying how inspired they were. Even though I had no time, I began to blog regularly about my experiences.

1.8. One of the neighborhoods devastated by the fire.

Months before, I'd committed to coaching people who would be attending a New Year's retreat that Christine and I offer each year. The next teleclass was scheduled for the exact time I was driving back. With no home and no office, how could I even think about honoring this commitment?

1.9. Treasures in the ashes.

Then I remembered that I had logged on to the Wi-Fi network at a hotel in Mill Valley a couple of weeks earlier while attending a seminar taught by my friend John Gray, who wrote the bestseller *Men Are from Mars, Women Are from Venus*. Mill Valley was on my way home, and the Wi-Fi password was probably still good. I pulled into the hotel parking lot and logged on. Sitting near the pool, I talked to the participants on my laptop.

It was hard to pay attention, to listen to the life visions of the people I was counseling, and not share the cataclysmic loss we had just suffered. I focused on their stories and no one guessed that I was teaching the class on a poolside deck instead of my Mark West office 2 days after losing my home.

When I got to Vancouver to deliver my keynote speech, no one wanted to see the 196 PowerPoint slides I had carefully prepared 2 months earlier. All they wanted to hear about was the fire. So I changed my keynote title to

"Through the Fire" and shared the raw reality of what Christine and I had just come through.

When I walked up on stage, 200 pairs of eyeballs turned toward me. Many of those attending were mental health professionals specializing in treating psychological trauma. I could see them silently diagnosing me to determine if I was plumbing the depths of denial or whether my light-hearted demeanor was real.

72 Hours

10/17/2017 03:23 p.m. ET - Huffington Post - Santa Rosa Fire Blog Post 2

It's funny how the mind plays tricks on you. Seventy-two hours after escaping from the fire that consumed our house, I'm packing for a trip. For a year, I've been booked to deliver the closing keynote at the Canadian Energy Psychology conference in Vancouver this coming weekend, followed by a 3-day trauma training. With family and friends taking care of so much, I'm now able to leave, only a day later than planned.

I have a routine before my many trips, workshops, and keynote speeches. I pack exactly the same things. The list has been honed so carefully over the years that I can now do a 6-week European teaching trip with only a carry-on bag.

Now my mind keeps thinking thoughts like, "I need to pack the soft black microfiber shirts I wear during healing sessions. Did I pick up my pants from the dry cleaners? Will my favorite silver jacket crush if I don't carry it on board the plane? I must put my headphones in my pocket."

Then I realize that I don't have a silver jacket. Or any shirts, pants, or headphones. I don't even have a suitcase. Even though it's been 3 days since the fire, it's taking a long time to adjust.

Just 72 hours ago, I didn't possess a pair of socks. When we ran from the house, fire all around us, all I grabbed was my phone, my laptop, and Christine's hand. No power cords, no toiletries, no treasured possessions.

Today I look through my old To-Do list on my laptop and laugh. The things that seemed so high a priority 72 hours ago seem trivial today.

I think occasionally about the possessions we lost. The 1861 edition of the complete works of Sir Walter Scott, published by Adam and

1.10. Christine with gifts.

Charles Black, Edinburgh. All 12 volumes. My great-great-grandfather's tintype photos from 1864. All the watercolors I painted. The 1,200 neatly organized folders of art lessons that made up Christine's art business. My red 1974 Jensen-Healey.

I don't miss any of it; life is infinitely more precious.

Even though the fires are still raging in some parts of the county, we see people picking themselves up and starting over. The folks next door to where we are staying started a remodeling project yesterday.

My heart is overflowing with gratitude this morning. I lived through one of the worst disasters in my area. Christine and I escaped with seconds to spare. I have a wonderful wife and family. I wake up in love every day. I feel guided by the Great Spirit every moment. The day after the fire Christine and I meditated for a long hour and we re-visioned the fire as an opening for the Universe to bring wonderful new things into our lives. Life is precious, and fire or no fire, we can choose to revel in its sweetness every day.

Barnett Bain, producer of the Academy Award–winning movie "What Dreams May Come," is one of the hundreds of people who've written encouraging emails to me. I wrote back and thanked him, saying, "We are whole in Spirit, which is the ultimate reality."

My friend and collaborator David Feinstein, a clinical psychologist who has authored several key textbooks in the field, gave me a whimsical introduction. "Either Dawson Church is a complete fraud," he said, "or he's an incredibly resilient human being and these methods we teach in our profession really work. I'll leave it to you to decide."

At the end of my talk, people jumped to their feet and gave me a long and enthusiastic standing ovation. The conference gave me an award for my scientific contributions to the healing field.

The week of training that followed the conference was poignant. Everyone knew I had just come through the fire in a very literal sense and people were incredibly kind. One of the primary methods I teach, Emotional Freedom Techniques (EFT), uses tapping on acupuncture points to relieve psychological trauma. I had done a lot of tapping since the fire, and I did a whole lot more on that trip.

In New York, one of our certified expert practitioners did several hours of trauma release work with me. The image that was burned in my brain was that 10-yard-high, candle-shaped flame. It took that plus more hours of treatment before I was emotionally neutral when recalling the image.

As I talked to my beloved wife on the phone every day, I could hear in her voice that she was suffering, so I cut the trip short. I canceled the last few workshops. I returned to California to find her at the center of a nexus of love, as her children and friends took care of her every physical need. But she needed me emotionally every day.

The Blessing of Small Things

10/24/2017 06:29 P.M. ET - Huffington Post - Santa Rosa Fire Blog Post 6

I'm wearing an old and battered pair of glasses. They're extremely precious to me.

Before the fire, eyeglasses were an expendable commodity. I ordered them in sets of three. That's because I knew that no pair would last long. I would sit on them, drop them, or leave them in places I visited. Each year, I obliterated several pairs that way.

Since the fire, only one pair remains, the throwaway eyeglasses that happened to be clipped to the visor of the car we escaped in. Suddenly, a thing that was of negligible value to me has become extremely valuable. Because of these scratched-up eyeglasses, I can see. I'm so grateful for that privilege. The fire has reordered my priorities.

Fingernail clippers. I always have a pair in my travel kit and a second in the top drawer in my bathroom. I've had that arrangement for so long it's slipped below the level of consciousness. Now I'm here in Vancouver missing my clippers.

I make a mental note to buy some at the first drugstore I pass, but every day I forget, because it's been a decade since I had clippers on my shopping list.

Every morning, when I pull my socks over my feet, I give thanks. Before the fire, I never gave a single conscious thought to any of thousands of times I picked up a sock. But losing them all in the fire, and being without them for 2 days afterward, has made me mindful of the blessing of soft warm fuzzy socks.

Imagine if we lived our whole lives mindful of the blessing of small things. Imagine if appreciating the tiniest of items, from toothbrushes to spoons to headphones, was our routine approach to life.

I have resolved to live my life from here on out taking nothing for granted. Yesterday I walked down the sidewalk, giving thanks for each breath. For the crisp British Columbia air entering my lungs. For the shoes on my feet. For the ability to walk. For the leaves falling from the trees. For the rain falling softly on my umbrella.

One of the many friends who's resonated with these posts I'm writing about the fire sent this poem by the medieval Persian poet Rumi called "The Guest House." It reminds us that the disasters of our lives can bring blessings, as long as we surrender and give thanks.

This being human is a guest house.

Every morning a new arrival.

A joy, a depression, a meanness,
some momentary
awareness comes
as an unexpected visitor.

Welcome and entertain them all!

Even if they are a crowd of sorrows, who violently sweep your house
empty of its furniture, still, treat each guest honorably.

He may be clearing you out for some new delight.
The dark thought, the shame, the malice,

meet them at the door laughing and invite them in.

Be grateful for whatever comes,
because each has been sent as a guide from beyond.

My wish for you is that you're able to see the problems in your life as "guides from beyond," summoning you to a consciously lived life. That you're able to surrender and give thanks for the miracle of simply being alive each moment. That you appreciate each breath, each sock, each pair of eyeglasses, and every other small thing in your life.

It took a firestorm in which I lost every material thing to wake me up to the preciousness of them all. My prayer for you is that it doesn't take a disaster to remind you to savor the magic of each moment. That you wake up grateful, give thanks with every breath, and go to sleep each night bathed in the wonder of a life mindfully lived.

Christine had made several visits to the Mark West property, and for the first time we went there together. We sifted through the ashes and were amazed. The fire had been so hot that the aluminum wheels on my Jensen-Healeys had melted and were just puddles of molten metal. A friend suggested I retrieve the famous Flying Lady emblem that had graced the hood of my Rolls-Royce. But the fire had melted the entire front of the car and the Flying Lady was no more. The windshield had melted. That showed that the fire had reached temperatures of over 2,500 degrees, the melting point of glass.

1.11. Formerly the aluminum wheel of a classic Jensen-Healey sports car.

We collected a few souvenirs, like shards of my grandmother's Royal Doulton china, but it was hard to distinguish individual objects in the rough gray rubble.

Surprisingly, a statue of the Buddha had survived intact. It sat in the ashes of our office building surrounded by debris. Calm and serene, it was a reminder of the permanence of intangible values. I wrote this blog about it in the *Huffington Post*.

The Saint in the Ashes

10/24/2017 06:42 P.M. ET - Huffington Post - Santa Rosa Fire Blog Post 12

Today was the first day my wife, Christine, was able to get to the property. She went with Heather, our wonderful EFT Universe operations manager who lives close by. Ray, Heather's husband, brought tools to sift through the ashes.

Christine said it looked like a blast zone. The emergency personnel at the scene told her the house had been reduced to ash within 5 minutes of the fire reaching us.

1.12. The saint in the ashes.

There wasn't much left to sift through. She found a few keys, but the metal had been twisted by the heat. Ceramics and sculptures had survived. The most poignant image was of this Buddha sitting in the ashes.

To me it represents the central truth of my experience in the fire: While material things may come and go, the core of eternal energy that is the truth of our being cannot be burned.

The important things—love, connection, compassion, awareness, trust, faith—cannot be destroyed. When everything around them is burned away by the fires of life, their outlines stand out more boldly.

When Christine and I talked on the phone after her visit to the property, we reflected on all of our blessings. We talked about the stuff we'd lost, and we also talked about the great future we know we'll create together.

The message of the Buddha is that the sooner we release the past, the sooner we can embrace a new and positive future. While there were many precious possessions lost in the fire, the possessions in our hearts are the ultimate treasure. With them, we will always feel rich and blessed, whatever challenges we face in the outer world.

RE-CREATING THE OLD

Christine and I were still acting in disjointed ways. The insurance company sent us $5,000 to cover our initial expenses, and I went on a tool-buying spree. I had worked in construction during and after college at Baylor University in Texas, so I have the skills to repair homes and cars. Soon one corner of our hotel room was piled with hammers and screwdrivers and saws and levels, as well as the tools to repair classic cars.

Consciously, I knew I no longer had a house. Or classic cars. The message hadn't yet reached my subconscious mind, and I was buying tools to repair items that no longer existed. It takes the psyche a surprisingly long time to adjust to radically new realities.

The way insurance works predisposes you to re-creating the old. If you buy exactly the same refrigerator or toaster you had before, the insurance company reimburses you for "replacement cost," the full value of the new appliance.

But if you don't replace a possession, they pay you only the "depreciated value," about what it's worth at a flea market. The whole system steers you inexorably toward re-creating a new life identical to your old life. Buying the same possessions, filling the blanks of your new life with the puzzle pieces of the past.

One day about 2 months after the fire, Christine and I looked at each other like people waking from a dream. We said to each other, "We don't want our old lives back. We don't want all our old stuff. We want to create anew. We want to build the fresh, living reality our hearts now desire, not re-create the old reality we used to have." So we stopped buying stuff, other than a few select items that we knew we truly wanted. We savored each purchase. We gave thanks for our new possessions—consciously and deliberately chosen, one by one.

We were grateful to Julia and Tyler for letting us stay in their apartment, and we later moved in with a brilliant artist friend and her husband. But we now faced the challenge of finding a permanent place to stay.

The property market had been tight before the fire. The inventory of new homes was sparse and prices were sky high. Rentals were almost nonexistent, and renters competed furiously for the few available.

Then 5,300 homes went up in smoke. It was impossible for insurance companies to find anything local for those displaced. They rented homes for people in Sacramento or San Francisco, a hundred miles away. Many were forced to live in trailers on their burned-out land, or park in the driveways of relatives. Where were we going to live?

THE VOICE

"Call Marilyn," Christine said to me one day. Marilyn is a friend of ours who is the former president of the Institute of Noetic Sciences, and she lived at the time in Petaluma, the place we had most wanted to relocate before the fire.

"Darling, I am beyond overwhelmed, and I don't have time to call Marilyn," I responded.

A few days later, in her sweet way, Christine again gently suggested I call Marilyn. When she speaks in that tone of voice, I know from long experience that she is "hooked up." Her voice changes slightly, and I can feel the angels speaking through her. I called Marilyn.

Marilyn was intrigued. She said, "I've been doing contract work in Silicon Valley and my husband works at Tesla. It's a long drive every week. If your insurance company will rent my house for you to live in, that will give me the money to rent the place I really want in Mountain View!"

With a single phone call, we had a dream house in exactly the place we most wanted to live.

THE FOUR SEASONS

We loved Marilyn's house. A gate in the backyard fence led to a huge open-space preserve. Two blocks away was a tennis court. Unlike Mark West Springs, the neighborhood was flat and level enough for us to ride our bikes right out the front door. Near the tennis court was a large county park with miles of mountain biking and hiking trails. I could paddleboard in the Petaluma River, a 5-minute drive away.

The house was twice the size of our old house, with many luxury features we'd previously only dreamed of. One day, with a shock, we realized that Marilyn's house was almost exactly what we had posted on our vision board as our ideal house. The universe had provided, though in a very unexpected way.

A friend and business colleague offered to let us use his spare office free of charge, and Heather and our core team packed into the space every Tuesday and Thursday, working from home the other days. We began to get our business, EFT Universe, as well as our nonprofit, the National Institute for Integrative Healthcare (NIIH), up to speed again.

The process took many months and was full of challenges. Just one was that the post office had trouble with mail forwarding for about 6 months, and all the checks we got in the mail were weeks late or never arrived at all. It was a financially and emotionally challenging time.

PAST AND PRESENT DREAMS

Once Christine and I became conscious that we were creating a new life for ourselves, and we could build it out of our present dreams rather than the visions of the past, we made different choices. I adopted what I called the Four Seasons standard. We've enjoyed Four Seasons resorts in different parts of the world, and they're always beautifully furnished. I decided that I'd like our new home to look as beautiful as one of those resorts. I told Christine, "I only want things in here that would belong in the Four Seasons."

I kept forgetting. A friend offered us an old slipcovered couch. It wasn't too bad; it had only one hole where her pet rabbit Whitey had chewed through the fabric. Because we had no couch, I accepted it and made arrangements to have it picked up in a borrowed truck. Christine tactfully asked, "Would that be in the Four Seasons?" I let the couch go.

My daughter, Rexana, was moving to Texas and gave us her basic household furnishings. We were in no hurry to fill up our lives with stuff, and we wanted the perfect décor in our new space.

We decided to buy the bed of our dreams, no expense spared, because we knew we'd be sleeping in it for a decade to come. We went to many mattress stores and lay down on many beds. Our friend Jane sells mattresses for children on her Healthy Child website, and she told us that there are many harmful chemicals that can be used in the manufacture of commercial mattresses.

Could that really be true? I looked up the scientific research on the chemicals that can legally be used in mattresses. I was shocked to find that Jane was right. We went all organic. For the base, we picked a split foundation that can be raised and lowered with a remote control.

Eventually, the bed arrived and it was everything we'd always wanted. But we realized after a few days that it was the perfect bed—for our old house. We'd again unconsciously created our future out of our past.

I was equally disjointed when looking for vehicles. I'd purchased several at auction previously, so I went online and bought a Toyota Prius and an old RV at an online auction. But it transpired that the auction company was headquartered in Oregon and title had to be transferred from the auction lot in California to Oregon for them to sign the title over to me—then transferred back to California. This Byzantine process took several weeks and I was without a car.

Christine said, "You're a car guy, why don't you put the word out to your Car Club friends?" I did, and 24 hours later I was driving a nice Ford F250 truck courtesy of a member of the Wine Country Car Club. He offered me either that or his 1924 Rolls-Royce, and while I loved the Rolls, it didn't work for moving couches.

12:45 A.M.

11/13/2017 05:31 P.M. ET - Huffington Post - Santa Rosa Fire Blog Post 11

It's 12:45 A.M. and I'm wide awake.

I've been wide awake at that same time every night for the previous week. I can't get to sleep for at least 2 hours, and then I toss and turn uneasily till dawn.

I can't figure out why. I may have the occasional struggle with insomnia, but this is uncanny. Nothing I am able to do is able to calm my racing mind. I tap, I meditate, and I still wake up at exactly 12:45 A.M.

Finally, it hit me. That's when I woke up on October 9 with the realization that something was wrong, the night I looked out the window and saw a wildfire racing toward our house.

Now my body knows that something bad happened at 12:45 A.M., and it wakes me up with a surge of cortisol.

1.13. Walking through the ashes.

I performed a key study on the effects of EFT tapping on stress hormones. Our research team randomized people into three groups and tested their cortisol levels before and after therapy. One group got regular talk therapy, one group rested, and the third group tapped. Anxiety and depression went down twice as much in the tapping group, and cortisol declined significantly.

So I know that these techniques work and I know what a cortisol surge looks like. I remember the story of a particular man treated in the Veterans Stress Project, which I founded. He'd endured a mortar attack at 4:45 A.M. on his first day of deployment in Vietnam in 1968. When he came in for treatment, more than 40 years later, he still often woke up at 4:45 A.M.

That's a typical cortisol surge. Though it was adaptive for getting our ancestors out of danger in past epochs, when it keeps on repeating, it plays havoc with the biochemistry of today's humans.

Since I'm waking up at 12:45 and staying awake despite my best efforts, I decide to make friends with the pattern. As I lie awake, I focus on being mindful of all the happiness in my life. The fact that I survived the fire. That I have a loving wife, successful children, and a magnificent community. That I have deeply meaningful work that contributes to the healing of thousands of people each year.

Exactly a month after the fire, to the day, I woke up at 1:45, an hour later than usual. And went back to sleep quickly. That meant my body was becoming convinced by my mind. It was no longer repeating the story that death is imminent unless we're on full alert at 12:45. The same thing happened the following night.

That's a positive change!

It's important to love our bodies. So often when they don't behave, by getting sick or developing patterns like insomnia, we want the problems to go away. We ignore them, deny them, suppress them, get mad at them, or medicate them.

If instead we can strive to understand our bodies and accept them just the way they are, we open the door to healing. Carl Rogers, the great client-centered therapist of the 20th century, called this the paradox of growth: We need to love ourselves just the way we are, with all our problems and limitations. When we do that, we start to change.

When your body knows it will be listened to, it can speak quietly. A little rumble here. A slight pain there. We hear the message and take care of its needs.

When I teach live workshops, I often work with people who've been ignoring or even hating their bodies for many years. They aren't attuned to the body's messages. They aren't picking up those subtle signals.

When its soft communications are ignored, the body has to speak more loudly. The small pain might become arthritis. If ignored, it might become a full-fledged autoimmune disorder. So many people are at war with their bodies, trying to mute their messages with medication or addictive substances.

Growth begins with self-love. Healing begins with self-acceptance, even when circumstances seem unacceptable. Practicing self-love lowers our stress levels and opens our awareness to the potential of our lives. Through that window of possibility, the love, peace, and beauty of the universe can shine. Even at 12:45 A.M.

THE DESIGNER UNIVERSE

After a big loss like the fire, we humans naturally crave stability and certainty. The fearful parts of Christine's and my personality wanted to rebuild at Mark West Springs, or buy a new house in Petaluma, as soon as possible. Restock our furniture. Fill our garages with cars and our rooms with stuff. Our things confirm the fact that we're alive.

Every lamp, every chair, every vase, every cup represents a totem of stability. Of normalcy.

But every gap, every empty corner of the house, had the potential to remind us of loss, of the beautiful things we once had and didn't have now. Till we paused to reflect, and decided to acquire things consciously, we rushed to fill the void.

Yet this rush to give evidence of our survival in the form of material possessions carries a high price tag. It crowds out the space in which our highest and unexpected good might organically unfold.

So each time we felt the compulsive desire to buy something or make a big decision, we paused. We meditated till we felt comfortable in the mystery. Our friends were puzzled that we were content to have big empty spaces in Marilyn's house for many months. Yet we took the opportunity to let go of our compulsive need for security and allow the universe to surprise us in synchronous ways.

Christine signed up for a weekly painting group. One of the other members, who'd also lost her home in the fire, bought a new house, fully furnished. She didn't need some of the items she inherited from the previous owner and gave them to us. Among them were two gorgeous Persian rugs. They were a perfect match for our new color scheme. A designer friend visited and told us they were worth $20,000. As we trusted the universe, it came through again and again.

Intuitively, we held back from making any quick or firm commitments, like major purchases or long-term leases. As we meditated each day, we felt as though we were aligning with a benevolent universe. We felt that our local human vision for our future would be limited, and that the universe had a much bigger dream for us than the small dreams that our limited human minds were capable of conceiving. Imposing our limited desires, especially in the panicked aftermath of the fire, would leave no room for the organic unfoldment of our highest good. So we just went with the flow. We wanted to leave space for the synchronous possibilities that might arise.

Butter + Coffee

12/07/2017 03:00 P.M. ET - Huffington Post - Santa Rosa Fire Blog Post 12

Yum! Coffee blended with a couple of tablespoons of unsalted organic butter. If you haven't tried it yet, it probably sounds weird to you. If you have tried it, you might well become an addict, like I am. My friend "Bulletproof" Dave Asprey introduced it to the world a few years back.

A cup of butter coffee in the morning fills you up for hours. It's part of the new fad of "intermittent fasting," which means limiting your food intake to 8 hours of the day. Intermittent fasting can be as effective as fasts in which you eat nothing at all, and which are notoriously difficult to sustain. I've been enjoying butter coffee for a couple of years.

This week I made my first cup of butter coffee since the fire. It's been 2 months. Our lives have been so disrupted that even the morning ritual of making butter coffee, which requires only a blender, coffee maker, and fridge, became impossible.

As I ladled the butter into the coffee, my body swelled with an inordinate amount of satisfaction. Butter + coffee = normalcy. The morning ritual was a tiny symbol that my life was returning to normal.

It's hard to explain how far from normal life has been. You're dressing in the morning and you want to match a shirt with a pair of pants with the same color socks. You realize that the socks are in a friend's garage, half an hour away, while the pants are at the cleaner's.

You want to charge your Bluetooth speaker, and then it dawns on you that the cord is in a distant storeroom and the power supply is missing. Tasks that take minutes when all your possessions are in your house take hours when they're scattered among different locations. You get further and further behind, even as the urgency to catch up increases.

Your inbox is full of messages from friends wanting to help. People a thousand miles away are offering you shelter. It makes you feel loved, and you appreciate their concern. Yet responding to their emails consumes yet more of the time you need to accomplish the simplest of daily tasks and get your life back into some kind of order.

So the simple act of making my routine morning cup of butter coffee made my body feel warm all over. Two months after the fire, it was a symbol of normalcy in a world turned upside down. When those symbols are few and far between, you treasure them.

We go through our lives taking those normal routines for granted. Our houses, families, and possessions are all around us, and it's easy to think they'll be there forever.

They may not be. Savor them while they last! That little ritual you engage in each day might be far more precious than you understand.

THE WORK TRIPLES

In the normal course of a workweek, I have a lot on my plate. Several hundred people move through our Energy Psychology and EFT certification programs each year, and one of my great delights is mentoring their progress. This means giving them individual feedback on their case histories, as well as conducting regular group teleclasses.

I write extensively for blogs, books, and online programs. I travel often to teach workshops in the US, Canada, and Europe, and I'm sometimes away from home and office for weeks on end. I have over 200 radio, summit, and podcast interviews per year. Then there's the everyday business of running an organization.

The National Institute for Integrative Healthcare (NIIH) is an amazing organization. Our biggest single program is the Veterans Stress Project, which offers energy therapies to veterans with PTSD. We help thousands of veterans and their spouses every year.

The other branch of the NIIH is the Foundation for Epigenetic Medicine. It performs research, usually with me as one of the investigators. We've played a part in over 100 scientific studies, and several energy therapies are now evidence based, partly as a result of our efforts.

The NIIH is an all-volunteer organization in which nobody gets paid. I put in about 500 hours a year to further its goals.

With all this activity, it's rare that my workweek is less than 40 hours; often it's twice that.

None of this stopped after the fire. Nor did the need to pay our team members and freelancers and for all the infrastructure that reaches over a million people a year. I still had to make all the appearances I'd scheduled—since that's what brings in the funds to pay for it all.

But now on top of this I had to start putting my life back together. Finding a place to stay, setting up a household, pulling together every single little strand of life.

Till you lose it all, you don't realize how much stuff you need to live a normal life. It's in the background, and you take it for granted. Potato peeler. Corkscrew. Serving trays. Water pitcher. Salad bowls. Coffee maker. Blender. Oven mitts. Egg lifter. Paper towel holder. Spice rack. That's just in one room, the kitchen!

After a disaster, you have to restock every room in your house from scratch. This takes time, focus, money, and energy.

Insurance companies require you to list everything for which you're claiming a loss. If you have an event like a burglary or flood, you can show them receipts for what's gone and they'll reimburse you for it.

But after a fire, there's no paperwork. It's burned up. So we had to reconstruct the contents of each room, item by item, from memory, and write it all down.

Being the organized person I am, I did this using Google Sheets, which allowed Christine and me to collaborate with our children on the effort. But it still took many days over the course of many weeks to make the lists.

I read a story in our local paper about how difficult this task was for fire victims. It said that doing an inventory can take 20 hours. I laughed; at that point, it had taken us over 400 hours. Four hundred hours—that's 10 weeks' work. And we were still far from finished.

Tubbs Survivor Pulled from Wreckage Two Weeks after Fire

11/01/2017 06:51 P.M. ET - Huffington Post - Santa Rosa Fire Blog Post 9

It did not seem possible that he could have survived. Every other living thing in the path of the October 9 Tubbs Fire had either fled to safety—or perished.

No one even looked for him for days after the fire, because it seemed impossible to believe that any creature could have survived the firestorm.

Two weeks after the catastrophe, the authorities began letting homeowners visit their devastated properties. Christine returned to our property on Mark West Springs Road and found only devastation where our house had been.

Christine walked through the wreckage, our friends Heather and Ray there to support her. Our pond, where Christine had cared for our elegant collection of koi, looked like a toxic waste dump. Kind Ray had removed the floating corpses of the fish before Christine could see them. Tears came easily.

As she walked by the pond, Christine thought she saw movement, though nothing could be living in the scummy black water. The next day, our resourceful friend Rick arrived to help and waded fearlessly into the mess. He groped around on the bottom of the pond, and his hands touched a domed object.

He lifted his hands and out of the black water came . . . our turtle! Somehow he had survived in the toxic stew for 2 weeks.

We had never given him a name. He is a "red-eared slider," a common species known for its hardiness. He came to live in our pond about 7 years ago.

It's not difficult to get out of the pond if you have flippers rather than fins, and in the first year he escaped twice. Both times we found him a short distance away. After that, he concluded that the perils of freedom were greater than the safety of captivity, and he no longer strayed. The grass isn't always greener on the other side of the hill.

He used to swim around with the fish, but he demanded special treatment. When we sprinkled food in the pond for the koi, he'd ignore the feeding frenzy. Instead, he'd glare at us with his tiny eyes until we knelt down and hand-fed him. He wanted love and connection, not just food.

After the fire we decided to christen him Mr. Tubbs. He was adopted by a friend with a bigger pond right near a stream.

Celebrate any thread that connects you to love and hope, no matter how fragile. We're familiar with the people and things around us and assume they're permanent. Like Christine and I before the fire, you have the illusion that the things you have and the people with you today will be there tomorrow.

Yet they could vanish in any one of a thousand ways.

1.14. Tubbs the turtle.

Treat them as though they might be gone tomorrow. Treat every moment of life as though it is precious, especially those elements that connect you to family, friends, spirit, and nature. You might never experience a catastrophe that takes them from you, and you will then have enjoyed them to the full, savoring their sweetness till the end of your days.

THE HEALING MANDALA

The New Year's retreat that I'd nurtured during the teleclass in Mill Valley filled up. We went to Hawaii as planned and spent a week leading a select group of 15 people through a profoundly transformative experience. It involved daily meditation, getting in touch with your higher power, and creating from the highest possible version of your identity. One of the exercises is drawing a mandala, and both Christine and I were inspired to use a bird as our theme. Leading other people through the visioning process gave us the opportunity to do the same.

One of the tangible things I lost in the fire was my collection of journals. I have kept a personal journal since I was 15 years old, and my whole shelf of journals was now gone. The day after the fire, I bought a new journal. I wrote down all the spiritual milestones of my life, summarizing

1.15. Healing mandalas. See full-color version in center of book.

decades of personal growth and transformation. Then in the next few pages I wrote down all of the material milestones of my life, like jobs and relationships, to anchor the new journal in the flow of emergence from the past.

Each day of the Hawaii retreat, we began with meditation before breakfast and then ate in silence. We picked inspirational cards, and did exercises to spark the creative process and get us in touch with our deepest inner guidance.

I've never liked the conventional New Year's ritual. Hanging out with drunk people and screaming and yelling as you watch the ball drop in Times Square on TV? No, thank you. So Christine and I created our own ritual with the group. At midnight we walked the labyrinth at the retreat center, silently holding our visions for the coming year. It was a full moon and a magical moment.

WE SUPPORT EACH OTHER AND THINGS WORK OUT

The last household item I replaced was a set of Japanese Mikasa cutlery. All the gifts Christine and I had received at our wedding many years before had been destroyed in the fire, and I missed a few of them. The day I replaced the Mikasa cutlery set, the house felt complete.

Four months later, we were organized enough to throw our first party. We invited our friends and filled the house with laughter, wine, food, and love. I stood up and shared all the miracles that had happened since the fire and our deep appreciation for their support. I felt we were at last back to normal.

1.16. Hawaii retreat.

One event at the party symbolized the 9 months since the fire. Bill and I were sitting next to each other at dinner, while Jane and Christine chatted at the far end of the table. As Bill enjoyed dessert and a glass of excellent wine, he sighed deeply. He pushed his chair back and relaxed into its embrace.

The back leg of the chair went over the side of the deck, and suddenly he was flying backward. The full weight of his body was catapulting his head toward the stone patio two feet below.

I grabbed Bill's forearm in a desperate attempt to stop his fall. He grabbed mine. I dug in my heels and willed myself to stay anchored to the deck, rather than being dragged over the side by the momentum of Bill's weight.

As our forearms joined, we felt like rocks. Bill's fall was arrested like he'd hit a wall. I pulled him toward me and his chair leg grated back up onto the deck.

What a metaphor for life! We support each other and things work out.

HEART-TO-HEART WITH TONY

The month of the party, my book *Mind to Matter* was officially published. It hit the #1 slot on the Amazon Neuroscience list the first day and stayed in the Top 10 the rest of the year. I'd finished the manuscript and sent it to my wonderful publisher Reid Tracy of Hay House just the month before the fire.

I spared no effort when writing and marketing the book. I read hundreds of scientific studies, then wrote up the evidence in a form that nonscientists could understand. It's all about the science of manifestation—how we can create material reality from our thoughts and live in synchronous harmony with the universe. The evidence convinced thousands of people to begin meditating regularly. Its ideas began to go viral, with people sharing it with friends and colleagues.

Several of my friends, like Jack Canfield, who co-authored the Chicken Soup series of books, know Tony Robbins, and for years I'd wanted to meet him in person. I'd thought of asking Jack for an introduction but hadn't taken any action.

Out of the blue, an email arrived in my inbox asking if I'd like to present at a Tony Robbins event. I excitedly said yes and asked Tony's event manager how he knew about my work. "He's been a fan of yours for years," she replied, to my surprise.

A few weeks later, a year after the fire, I found myself on a plane bound for Abu Dhabi in the Middle East to meet Tony and present EFT and meditation to his community. Mind to matter.

I enjoyed the huge bear hug Tony gave me when we met. He's even taller than I am, and the two of us stood like giants on the stage, embracing the audience in the energy of our passion for transformation. I tapped with his people and then led them all in a profound meditative experience. I flew home feeling as though meeting Tony had been an initiation—one in a long string of synchronous blessings, like the phone call to Marilyn, through which the universe affirmed the value of my life and work.

A week later, I led a workshop at Esalen Institute in Big Sur, California. On the last day, Christine and I packed our car and drove to the office to pick up a friend who was carpooling with us. I was impatient to leave, but we couldn't find our friend at the rendezvous point. Christine went to look for her while I sat waiting in the car.

1.17. High fives with Tony.

Twenty minutes later, I noticed a man walking by. I recognized him immediately: Tim Ferriss, the author of *The 4-Hour Workweek* and *Tools of Titans*. One of the top bloggers on the planet.

"Tim!" I called out. I introduced myself and we had a talk about EFT, which he had been wanting to investigate. The universe had just arranged for him to meet me, the author of the latest edition of *The EFT Manual*! And if Christine and I hadn't been late connecting with our friend, I would have left Esalen before Tim walked by. Synchronicity after synchronicity.

I emailed Reid Tracy with an outline for a new book on ecstatic brain states, and after the success of *Mind to Matter*, he and his team were excited about its potential. I began gathering the ideas I wanted to present. The outline began with a chapter on how meditation activates the happiness centers of the brain.

In subsequent chapters, I planned to show that as we use these parts of the brain daily in meditation, they grow, adding synapses and neurons. At the same time, the fear-processing centers of the brain, no longer being used, start to shrink. Over time, our brains literally remodel themselves, turning the software of mind into the hardware of brain.

This change in brain composition makes us resilient. When we have a setback in life, we have dense tissue in the happiness parts of the brain.

We have the neural hardware to bounce back quickly. The term "posttraumatic growth" refers to how navigating life's inevitable challenges and disappointments can actually make us stronger. I've written two books about psychological trauma and conducted several scientific studies with war veterans. I believe that with all the public focus on PTSD, the alternative possibility—that trauma can make us stronger—is often overlooked.

POSTTRAUMATIC GROWTH

Hay House encouraged me to tell my story of the fire as an example of posttraumatic growth. Sure, reading an expert's scientific explanations can be fascinating, but hearing from a person who's literally "gone through the fire" puts you face-to-face with reality. So the story of how I put my teachings into practice and used the fire as a springboard for posttraumatic growth became this first chapter.

We all have fires in our lives. They may burn up our marriages or destroy our relationships with our kids. They can sear our work lives and turn our careers to ashes. They may incinerate our wealth or retirement plans. No human being is exempt.

Traumatic experiences send many people into a spiral of fear from which they never recover. They become vigilant for possible threats, anxious about the future, and suspicious of joy. Nightmares, self-defeating beliefs, and intrusive thoughts haunt their minds.

Yet the same traumatic experiences can spark healthy reappraisal of one's old values. They can open up new horizons of possibility. Anchored in a relationship with a loving universe, we use even the most devastating loss as a springboard for growth. Practices like meditation and tapping give us access to our strengths, take us to a higher perspective, make meaning of disaster, and create even more profoundly wonderful lives than we had before.

I have a burning passion to awaken people to this possibility.

FROM 50 YEARS TO 50 SECONDS

In this book I share the discoveries of the 50-plus years I've spent investigating human potential since I first learned to meditate as a teenager.

I also—unusually—write about my own life. Normally I'm writing about science, and letting research tell the story. I always illustrate the science with examples of how real-life people apply it in their lives. Here in *Bliss Brain,* the real-life person whose experience I describe is sometimes me.

In Chapter 2, I'm going to show you why most people find it so hard to meditate. The difficulty has nothing to do with willpower or intention. It's simply due to the *design* of the human brain. When you understand this clearly, you'll be equipped to work around it.

Chapter 3 describes the ecstatic states that you can achieve in meditation. It examines the regions of the brain that you activate, and what each one does. It also lists the extensive health and cognitive benefits that you get from activating each of those regions.

In Chapter 4 you'll hear the story of my failed meditation experiences. I learned many different styles of meditation, but I could never establish a consistent practice. My breakthrough came from science. When I combined seven simple evidence-based practices together, I found a formula that *puts people into deep states automatically and involuntarily.* No effort required.

When I and my colleagues hook people up to EEGs and MRIs, we find that using these seven steps, even non-meditators get into profound states in less than 4 minutes. Sometimes in less than 50 seconds.

Historically, the secrets of these states have been available to only about 1% of the population. Thanks to science, they're now available to everyone.

Chapter 5 is about the seven neurochemicals of ecstasy. We'll learn how each one is like a drug that makes you feel good. But combine all seven together, and you have a potent formula that takes your brain into bliss. Meditation is the *only* way you get all seven at one time. The star of the show is a neurotransmitter called anandamide, aka "the bliss molecule."

When you trigger these ecstatic states daily, they change your brain. Chapter 6 is about the extensive brain remodeling that occurs in seasoned meditators. Stress circuits shrink and atrophy, while happiness networks grow. But you don't need to be an adept to trigger this rewiring. It begins the *very first week* you meditate effectively.

Chapter 7 is about posttraumatic growth, and how the brains of meditators make them resilient to the inevitable upsets of life. Medical crises and financial disasters included. It provides practical examples of how meditation can make you resilient even during global upheavals like the coronavirus panic and economic meltdown of 2020.

Chapter 8 delves into a fascinating question: Just how much can our brains change? The answer is: a lot. Reading the research on this question led me to a startling new scientific hypothesis, which might be changing the entire course of planetary history. It shows how this trend showed up as "caremongering" (rather than "fearmongering") during the crash of 2020.

The Afterword entices you to make these happy inner states your new reality, unlocking the full extent of your human potential.

Each chapter ends with Deepening Practices and an Extended Play Resources section. The Deepening Practices are simple activities you can use to apply the lessons of that chapter in your own life. The Extended Play Resources section links you to audio, video, and web resources that will enrich your experience of this book. It includes a companion meditation based on the theme of that chapter.

As we go on this journey together, you're going to learn all the best practices that I've learned in the past half century, and the fastest way to apply this science to your own emotional state. Use the Deepening Practices and companion audio meditations, and by the time you finish this book you're highly likely to have elevated your happiness to a whole new level as your brain starts remodeling itself.

ONE YEAR AFTER

Christine and I spent the 1-year anniversary of the fire having dinner with Julia and Tyler in our new Petaluma home. That same week, Lionel visited from New York and Rexana came from Texas to look at sites for her upcoming wedding. The previous weekend, we had gone camping and kayaking with the whole family including our grandchildren. Then I flew off to New York to teach another class.

The whole month seemed like a symmetrical celebration of the love and support that had flowed through our lives since the fire.

At my New York workshops, I had neurofeedback experts hook up participants to EEGs. Their brain waves showed remarkable changes. As I worked on this book, I read dozens of EEG and fMRI studies. I'd always known that meditation changes the brain, but new research was revealing much richer layers of information about the process.

I had dabbled in meditation from the age of 15 and became a daily meditator when I turned 45. I refined my meditation practice over the years, and when MRI studies of Tibetan monks were first published in the late 1990s, I became fascinated by the potential of meditation to produce large improvements in human health and longevity.

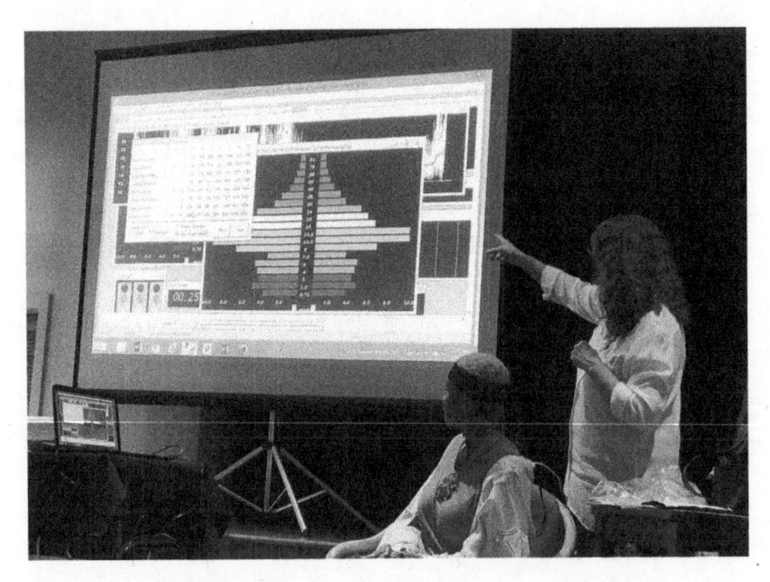

1.18. Measuring brain changes in workshop participants.
See full-color version in center of book.

Like so many people, my early attempts at meditation were unsuccessful. But in my midforties, to determine which methods were truly effective, I turned to science instead of relying only on ancient lore. This led me to develop a very simple but evidence-based form of meditation that people found easy to do. Everything in my life began to change when I began practicing it daily. Using EFT and meditation allowed Christine and me to recover from the trauma of the fire in record time.

For the whole first year after the fire, Christine and I started each day with meditation. As we sat and breathed, our worries and insecurities began to drop away. I found myself ascending in consciousness to an ecstatic state. I called it Bliss Brain.

Bliss Brain was our anchor point during the first chaotic months after the fire. It gave us a sense of well-being and a feeling of connection to a reality that extended far beyond our little local story. As we ascended in consciousness each morning, our perspective changed, and we saw our lives as part of the whole tapestry of being, bursting with fresh potential, rather than as a lonely island of uncertainty and tragedy. Here's where meditation took us right after the fire:

The Moment Everything Changed

10/19/2017 08:59 P.M. ET - Huffington Post - Santa Rosa Fire Blog Post 5

The first 48 hours after the fire I was dazed and confused. I couldn't figure out what to do next. Rumors abounded and hard information about what was going on in Sonoma County was scarce. People around me were fearful and anxious, and so was I.

Then something happened, after which everything was different.

My wife, Christine, and I woke up in Fort Ross, at the hotel we'd sought refuge in after escaping the fire that took our house. The night before, the sun had set in a red sky colored by ash and fire. We spent an uneasy night, haunted by dreams of driving through flames.

That morning, I said to Christine, "We need to do something urgently."

She looked at me curiously.

"We need to meditate."

We sat upright in bed and tried to enter that space. It was difficult. Images of yesterday's bloody sky kept popping into my mind. I realized how off-base my energy was.

The fire had burned through the rope tying the boat of my life to the anchor of my spiritual practice. I had been drifting in fear and uncertainty for hours on end. It was time to connect up again. To tune in to the Universe, that great eternal radio station broadcasting peace and serenity 24/7.

As we breathed and centered ourselves, the familiar energy of peace and calm washed over us. We were back in the heart of Great Spirit, the home that can never be damaged or destroyed.

We sat for the best part of an hour enjoying the depth of connection. We then turned to each other and began sharing our thoughts. Yes, there had been a fire. Yes, it had burned down our house. But here we were, safe and centered in the core of being.

We began to think about the possessions we'd lost. But now, sitting in the Heart of Spirit, we saw them differently. I remembered the four boxes of my mother's photographs stored in the garage. When she died 20 years before, she'd left mountains of possessions to sort through.

The bigger ones had been steadily disposed of. But who has time to sift through thousands of snapshots from the 1960s, most of them blurry and featuring people we didn't know? The four boxes had occupied an entire shelf in the garage.

Photos are heavy, and they pressed against the sides of the boxes. The cardboard sagged a bit more every year. The soggy shape of the disintegrating boxes mirrored the joylessness of the unappealing chore of sorting through them.

Now those four boxes were no longer there to reproach me with their misery every time I walked by. They were ash. What a relief!

In another of our garages was a massage chair. We'd purchased a new one the year before and stored the old one in the garage. On my To-Do list was: Sell old massage chair on Craigslist.

But a year had gone by and I'd never found the time. Now the chair was ash, and my obligation miraculously removed.

I owned a beautiful classic Rolls-Royce Silver Spirit. But I had so much money tied up in it I could not afford a practical new car. I had made the occasional attempt to sell it, but now it was burnt to a crisp. The insurance money would now pay for that new vehicle.

Christine and I had been talking about moving to Petaluma for a few years. But with our life savings tied up in the Mark West house, it was a pipe dream. Now, with nothing to go back to, a move was inevitable.

We began to list the blessings of the fire.

In psychology, this is called a cognitive shift. Same picture, different frame. Same fire, different meaning.

As we recited our blessings after meditating, our mood shifted. We celebrated being alive, with infinite possibilities opening up to us. We began to feel cheerful and happy. We joked with one another. Just 48 hours after we'd "lost" everything!

After meditation, the fire meant freedom, not loss. It meant the burning away of the old, not the loss of a lifetime of treasures. It meant an opening to a new and better life, not sinking into the tragedy of the old one.

Meditation made the difference.

That day, less than 48 hours after the fire, meditation changed my whole outlook on life. I went from lost and confused to confident and happy. I went from purposeless and fearful to balanced and joyful. I reconnected with the vibrant, resourceful version of myself.

A few days later, I wrote in my journal: "I feel incredibly happy. Loved and protected by Spirit in each moment. Blessed with my community, friends, kids, and Christine. The fire seems like no loss at all—the house and possessions seem trivial by comparison.

"We don't know how everything will turn out. But we don't need to. We can just go with the flow. When you're anchored in Spirit, you're secure. You aren't tethered by material possessions; they're not what gives your life meaning.

"Meaning comes from connection with Spirit. Spirit is not a vague metaphysical abstraction, it's the foundation of reality! I choose to live there every day. It's profound to realize that NOTHING can take your happiness away."

DEEPENING PRACTICES

Here are practices you can do this week to integrate the information in this chapter into your life:

- **Journal Purchase:** Buy a brand-new personal journal. Don't order it online; instead go into a store. Pick up the journals and fondle them. Find one that feels just right to you. Even if you already own a journal, purchase a new one. Make the purchase a symbolic act of starting fresh. Choose carefully and forget about the cost. This new journal represents a fresh start in your life. Use it to record the insights you have as you read this book and use the free meditations that accompany it.
- **Gratitude Practice:** Spend 5 minutes each morning or evening writing in your journal at least five things you're grateful for today. Want a double dose? Do this *both* morning *and* evening. You can't overdose on gratitude!
- **Disasters That Never Happened:** Spend 20 to 30 minutes during a quiet time this week reflecting on all the things that could have gone wrong in your life but didn't. Perhaps a friend is going through a miserable divorce and you're

happily married. Appreciate the disaster that your marriage might have been.

- **Appreciate all the near misses in each field of life.** This includes career (e.g., "When the company downsized, I was sure I'd lose my job, but instead I was promoted."), money (e.g., "I sold my stock in that company just before it crashed."), relationships (e.g., "I might be in Suzie's shoes right now because I dated Gus before she did."), and health (e.g., "This winter I didn't get the flu when everyone else did.").
- **Meditation Practice:** There is a companion meditation to every chapter in this book. They're all 15 minutes or less. Listen to the one below every morning, before you do anything else. Get into the habit of giving yourself this gift of quiet and centered time every morning. It can change your life.

EXTENDED PLAY RESOURCES

The Extended Play version of this chapter includes:

- Dawson Guided EcoMeditation: Activating Bliss Brain
- The Seven Steps of EcoMeditation
- Audio Program on How Our Brains Transform Energy into Matter
- Video of Dawson Keynote Speech on Genetics and Resilience
- Calendar of Upcoming Workshops and Live Events

Get the extended play resources at: BlissBrainBook.com/1.

CHAPTER 2

FLOW

A TRAVELER'S GUIDE TO BLISS BRAIN

What is the experience of Bliss Brain like?

As I sit here in meditation this morning, I'm going to attempt to describe it, like a traveler reporting back to his friends from a distant country.

First, I close my eyes and go through the seven steps of EcoMeditation listed in the Extended Play Resources of Chapter 1. I use acupressure tapping to release any stress in my body, and any mental or emotional obstacles to complete inner peace. I relax my tongue on the floor of my mouth. I breathe through my heart, and slow my breathing down to 6-second inbreaths and outbreaths. I imagine my favorite beach and my favorite people playing on it, and I send a beam of heart energy to the scene. I picture a big empty space behind my eyes.

I can feel my body, and I'm dimly aware of my surroundings. But most of my consciousness is focused on the experience I am having at the level of pure awareness.

Waves of bliss begin pouring through my brain and body. Occasionally, I shake or sway slightly as another wave of ecstasy hits. I focus intently on the space in the center of my being. Between my eyebrows, my forehead and skull tingle at the spot where the connection with this elevated state of consciousness anchors itself most strongly in my physical body.

It's easy to slip out of this place. Just one stray thought will do it. That thought will lead to another and before I know it, my awareness has slipped out of Bliss Brain. I find myself mentally composing a long email to my marketing manager. *I know you want an opt-in page, but I think a full sales page would explain our program better.* Reliving an ethical dispute with a colleague. *Recommending a hip replacement for such an elderly patient was going too far.*

Random scenes from the movie I watched last night flit through my mind. *How accurate was Chris Pine's Scottish accent when he played King Robert of Scotland?* Fragments of a bizarre dream pop into awareness—*wearing only a feather loincloth, I am frying eggs for Tony Robbins in the San Francisco airport.* Ideas about how to present the data in a scientific paper I am writing. *Should I be using a t-test or ANOVA?* That attack of arthritis earlier this week. *Will I need a painful knee replacement someday?* I'm past the deadline for a keynote speech I need to write. *Will the conference organizer send me yet another snarky email?* And a million other distractions.

Whenever that happens, I return my attention back to center. It's like tuning to a radio station. I can easily lose the signal and let the dial wander to a different station, one filled with anxiety and stress.

But I know what the bliss station feels like. I know the music it plays and how my body feels when I'm absorbed in it. Because I've been to the center so many times, I can usually find that station just a few minutes after I close my eyes.

So I tune in there again now. I feel an immediate expansiveness in my consciousness, a sense of connection with the entire universe. I feel a sense of welcome, as though I've come home. I'm living at the address in consciousness where perfect well-being is the only reality.

As I retune myself to center, another wave of bliss floods through my brain, mind, and body. I feel my consciousness lift out of my normal state, like a balloon rising in the wind, to meet and merge with a consciousness so vast and expansive that it has no end.

I know that this is the same intelligence that runs the universe in such perfect order. It has a sense of rightness to it that all the cells in my body respond to. Every cell knows it's come home, that it's connected to the universal consciousness with which my mind has merged. The local reality field of my mind and body surrenders to union with the great nonlocal reality field of the universe.

There is no room in this consciousness for worry, doubt, or fear. The anxious thoughts with which I began the meditation session are now left far behind me, as the balloon soars high above the world of ordinary local reality.

My breath slows and deepens. Every breath is a connection with that great universal consciousness. Every inbreath flows out of that consciousness, while every outbreath flows into that consciousness. A warm feeling of well-being floods my body. Though the cool morning air felt chilly when I began the meditation, my body is now infused with the glow of connection.

As I center myself again and again, I notice an intense glistening silver-white vortex of light above my head. I drift up through the portal. I find myself in a level of undifferentiated light. I look down at my mind, and it is flooded with that same white light. I am in Bliss Brain.

Everything dissolves into the light. There's no body, no me, no mind, no universe. Only the light. The light simply is. It has no beginning and no end. It stretches to infinity. It's all there is; there's nothing else in this real world of light other than the light. I lose myself in oneness with the light.

2.1. Entering Bliss Brain.

There's a tingling pressure in the center of my forehead where the connection to the light tunnel is strongest. Angelic music echoes in my brain, sound adding itself to light. My body sways spontaneously from side to side. Muscles twitch as huge jolts of energy surge through. Every sensation is washed away as the blissful light of oneness streams through my body. In the back of my mouth, I taste drops of sweet ambrosia. My heart is filled with a vast sense of calm.

ANCHORED IN BLISS

I open my eyes and gaze softly at the room in front of me. Then I close them again. Once I'm anchored in bliss, it doesn't matter whether my eyes are open or closed. I can maintain this expansive state of awareness either way. My intention is to take this awareness into my workday after meditation. I don't want to think or act from anywhere else.

After a while, I look down again at my body sitting in the chair. I realize I've been drifting, one with the light, basking in bliss, for a long time. My heart fills with joy and my eyes fill with tears, as I'm overwhelmed with gratitude. For my life, exactly the way it is. For every detail of what is. For everything that will happen in the future, no matter what it might be. I give thanks for all of it. I connect with everyone who's meditating at this same time anywhere in the world.

I open my eyes and look at the sunlight outside the room I'm meditating in. I'm aware of both time and space again. The forgotten cup of coffee in my hands is completely cold.

Tears of gratitude flow down my cheeks. I look at my cup. The words printed on it read: *Find Joy in the Journey.* After we moved into the new house that replaced the one destroyed by fire, I went on a hunt for mugs printed with inspirational words; those bequeathed by the fire victims' shelter with captions like *Construction Equipment Dealer's Association* and *My Dang Dog Also Drinks from My Cup* didn't echo the energy of the meditative state.

I feel grateful for everything. My hands, with which I hold the coffee cup. My feet, with which I can walk. My breath, bringing life to my cells. My connection to the universe. The wonderful people in my life.

I close my eyes and I am immediately in the light once again. I open them and the light remains.

In a trance, I stand up and get a fresh cup of coffee.

My wife has woken up and she comes into the room to get her morning cup of coffee. We embrace wordlessly. I bury my face in her hair and am enraptured by its scent.

We gaze deeply into each other's eyes and say nothing as she sits down to meditate too. When I close my eyes again, I'm back in Bliss Brain.

MOVING INTO THE DAY

I realize I have an appointment coming up in less than an hour and I need to prepare—to decide on today's priorities, do the most important things first, and then look at my inbox. To carry the energy of this first hour of meditation into the mundane affairs of the day. The process of communion with the universe is complete.

In my Bliss Brain state, I send love to everyone and everything I will encounter in my workday. All the amazing people on my work teams. Those I will email or phone today. Those collaborating on research or training projects.

I tune in to all the people I'll see and meet and interact with in the future, through my blogs, teleclasses, radio shows, podcasts, emails, social media, and keynotes. I feel a connection with everyone in my future and everyone in my past. I feel love flowing from my heart to all of them.

I carry within me the indelible imprint of that time spent in communion with the infinite. I know that it will infuse my whole day, elevating my mind to a level at which it would never be capable of functioning unless I had centered myself at the start of the day.

The insights and ideas that arise in and after meditation are usually at a level of brilliance far above that of which I am capable in my ordinary waking consciousness. From this elevated perspective I'm making connections in a way that my ordinary consciousness cannot match.

I know I will find solutions, solve problems, and experience breakthroughs that I would never have had were my daily activities not infused with the wisdom, creativity, clarity, and joy of Bliss Brain. This produces a fundamentally different life from one lived at the level of ordinary consciousness.

I lived at that address for a long time before I discovered the ecstasy of connection with the infinite. At that level of ordinary reality, I believed my fears were real. I believed that my limitations were objective facts. I believed that who I was today was determined by my past experiences. My mind was trapped in a small subset of possibilities.

Now that I know that the expansive state is possible, and that I can reach it in meditation every day, I see limitless possibilities. I'm no longer stuck in that small local mind that sees problems as real and limitations as facts. When I move into Bliss Brain, I see vistas of possibility in which those problems and limitations cease to exist. They are only real at that limited level of mind, and they disappear when you consciously choose to ascend your awareness to the level of infinite nonlocal mind.

You then bring the solutions and possibilities of that level back down to your daily walk through life. This creates a completely different experience than a life trapped in the confinement of local mind.

I meditate each morning—rarely missing a morning—so meditation is the anchor point of my days. When I look back on the many years behind me, all that seems to matter is each morning's meditation. My time in the real world. My time in ecstasy. It's the stitching that holds the whole fabric of life together. This feeling of expansive connection is the most important experience of my life, a light that illuminates every lesser experience.

2.2. When we start the day in Bliss Brain, it influences our everyday reality.

Common Characteristics of Mystical Experience

My experience in meditation is not unique; it's common to meditators throughout history. A 14th-century Tibetan mystic, writing about his experience deep in meditation, described it as:

. . . a state of bare, transparent awareness;

Effortless and brilliantly vivid, a state of relaxed, rootless wisdom;

Fixation free and crystal clear, a state without the slightest reference point;

Spacious empty clarity, a state wide-open and unconfined; the senses unfettered . . .

The mystical experience isn't the property of Buddhists or Catholics or Taoists or Hindus. It's the common root of all religions. The great spiritual teachers entered these experiential states, and when they "came back from the mountaintop," described them to their followers in the idiom of their cultures.

Theologian Huston Smith, author of the best-selling textbook *The World's Religions,* called mysticism the pinnacle of all religions. True believers may dispute the fine points of theology, but there is no disagreement among the mystics, from whatever faith they hail, because they have shared the same primary experience. These experiences are *spiritual* rather than *religious.* Research shows that people often become much more spiritual after mystical breakthroughs, while interest in formal religious affiliation declines. Fourteenth-century Sufi mystic Hafez exclaimed:

Am I a Christian, a Hindu, a Muslim, a Buddhist, or a Jew? I do not know . . . for Truth has set fire to these words.

The adept no longer identifies with a particular religion, but with the source of all religions: the primary mystical experience. Nor with a theology, but with the source of all theology: intimate knowledge of the universe itself.

The first serious researcher to examine the characteristics of this ecstatic state was Andrew Greeley. In a study of 1,467 people, he drew parallels between mystical and ecstatic nonordinary states. Greely found these elements common to transcendent experience:

- Feeling of deep and profound peace
- Certainty that all things will work out for the good
- Sense of my own need to contribute to others
- Conviction that love is at the center of everything
- Sense of joy and laughter
- An experience of great emotional intensity
- Great increase in my understanding and knowledge
- Sense of the unity of everything and my own part in it
- Sense of new life or living in the world
- Confidence in my own personal survival
- Feeling that I couldn't possibly describe what was happening to me
- The sense that all the universe is alive
- The sensation that my personality had been taken over by something much more powerful than I am

As we escape the subjective self and rise above our suffering to view our experience objectively, we abandon the limitations of our local minds in the embrace of nonlocal consciousness.

Later researchers built on Greeley's initial findings. They found seven commonalities, including a sense of unity, enlightenment, awe, and bliss. The sensory vividness of the "enlightenment" experience exceeded that of everyday life. In his 1954 classic *The Doors of Perception*, philosopher Aldous Huxley called this "the sacramental vision of reality."

All mystics have similar experiences, whether they are Hindu *sadhus* begging as they wander the countryside, Buddhist monks isolating themselves in caves high in the Himalayas, or Christian nuns engaged in contemplative prayer.

Harvard University's first professor of psychology, William James, after his own transcendent experiences, observed in 1902 that "our normal waking consciousness . . . is but one special type of consciousness, whilst all about it, parted from it by the filmiest of screens, there lie potential forms of consciousness entirely different." He said that no account of the universe would be complete without accounting for these states.

These altered states are more than *subjective* experiences. Science now shows that they are *objective neurological states* as well. In his book *How Enlightenment Changes Your Brain*, neuroscience researcher Andrew Newberg calls this mystical ecstasy a "subjectively and neurologically real experience."

FALLING FROM GRACE

Meditation is called a *practice* for a reason: You practice it moment by moment.

Yesterday I had a perfect meditation, the one described above. Today's is less successful. As soon as I close my eyes, a thought rushes in to disturb the peace. Then another, and another.

Did I leave the kettle in the kitchen plugged in after I made my tea? I suspect that my colleague is saying nasty things about me behind my back. Is today the day the dry cleaning is ready for pickup? Elizabeth Gilbert is a much better writer than I am. When I was arguing with my father-in-law about his retirement plan last week, I could have made a much better point. Is it a leap year? When is my next performance review? Did the kids hear us making love last night?

I worry about each thought. Then I release it and return to Bliss Brain. I smile.

Then another thought rushes in to take the place of the thought I just evicted. I frown. I start eviction proceedings all over again.

I become frustrated at the constant invasion of my peaceful meditative state by thoughts and worries. There are days that it never stops and I silently scream, "God, give me a lobotomy."

Yet that frustration is just another worry to let go of. I love my mind because it's so inquisitive and curious, and thinking and worrying is what the mind does. Wanting it to change is like standing on the beach and wishing the waves to stop. It's a futile request that can only lead to more unhappiness.

What I can do is keep returning, persistently and deliberately, to Bliss Brain. I can maintain positive energy and good humor despite the unceasing activity of the mind. This practice trains the brain in the experience of serenity and to reorient itself to peace when disturbed.

This practice of deliberately flipping the switch back to Bliss Brain *may be more important* than attaining Bliss Brain itself.

We're accustomed to believing that what counts is attaining our goals. For the meditator, the goal could be holding a consistent state of bliss. Surely the saint who succeeds in attaining everlasting enlightenment is the pinnacle of perfection?

Perhaps not. Perhaps the knucklehead who falls from grace, then picks himself up and clambers toward it again, only to fall back into the pit and repeat the process, is doing something important. The *act of persistence* may be more essential to the meditator than the permanent attainment of Bliss Brain. The true hero may be she who persists, not she who wins.

So in your meditation practice, fail happily. Fail an infinite number of times. Falling from ecstasy isn't the problem; it's part of the process. Choosing to elevate your state back to Bliss Brain even if you only succeed in maintaining it for a second is *the* crucial component of a successful meditation practice.

THE MEDITATION CYCLE

An important study performed at Emory University looked at which brain regions are active during the falling stage as well as the meditation experience. The researchers found four distinct phases to the cycle. They called them Focus, Mind Wandering, Awareness, and Shift.

Focus is when attention is centered in the meditative state. Then **mind wandering** begins. We become **aware** that the mind is wandering, then **shift** ourselves back into focus.

Seeing meditation as this four-part cycle, rather than as a single ideal state, gives us a more realistic picture of what to expect in our meditation sessions. The Emory researchers emphasize that "Focus is inevitably interrupted by Mind Wandering (MW)." We then become aware that our attention has been hijacked and practice Shift. We shift our attention back to Focus on the meditative state. And so the cycle goes, time after time.

I used to envy those lucky few people who'd broken through to the other side, who had attained a permanent state of bliss, eternal focus uninterrupted by Mind Wandering. Who'd achieved permanent connection to the universe.

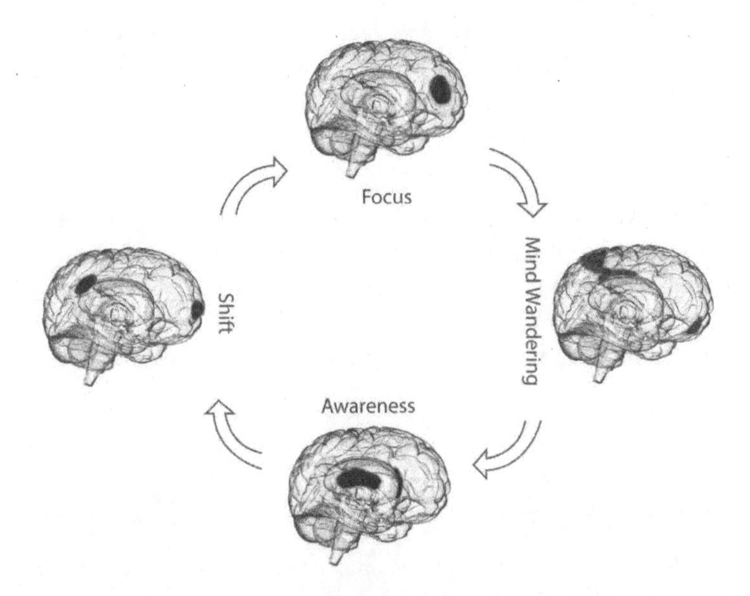

2.3. Four stages of the meditation cycle. The meditator (1) achieves Focus in the meditative state; (2) strays into Mind Wandering; (3) notices the deviation, which is Awareness; and then (4) consciously Shifts back.

Like the Buddha after enlightenment, sitting under the Bodhi tree. Like Eckhart Tolle, who just "woke up" one day while sitting on a park bench and stayed there ever after. Like Byron Katie, who woke up to transcendence in the midst of the deepest despair.

I used to compare them with plodders like myself, who after more than a half century of practice, still have to go through all four stages, getting sucked out of meditation by their wandering minds and then having to rescue themselves, time after time, even within a single hour's meditation.

But now I understand that it's the journey that counts, not the destination. That heroic journey back to Focus, even when it has to be repeated countless times, is what tilts the scale toward bliss. St. Francis de Sales (1567–1622) wrote, "If the heart wanders or is distracted, bring it back to the point quite gently . . . and even if you did nothing during the whole of your hour but bring your heart back . . . though it went away every time you brought it back, your hour would be very well-employed."

It's like flipping a switch that's programmed to be in the "off" position. You notice you're in the dark, and you flip the switch to "on." You're in the light (Focus).

But then the switch flips back again, because that's its default position (Mind Wandering). You again notice you're in the dark (Awareness). So you flip the switch again (Shift). You're in the light again (Focus). Discover you're in the dark, and you regard it as a signal to flip the switch back on again.

Do this often enough, and the ratio of dark versus light changes. In each meditation, you're gradually in Bliss Brain more than you're out. Each deliberate flip of the switch is a declaration to the universe that you're committed to being in the light. It isn't the length of time you're in the light that counts; it's the moments you're in the dark and flip the switch yet again.

The Cockroach Crawling over My Foot

After her divorce, Byron Katie sank into a severe depression. She became unable to leave the house and then unable even to leave her bedroom. Agoraphobic, paranoid, and suicidal, she turned to drugs and alcohol. She lived in this state for nearly a decade, ending up in a halfway house.

The other women in the house were afraid of her, so she was confined to the attic, where she slept on the floor. There she had a transcendent experience that changed her life completely.

"One morning I was asleep on the floor and I felt this thing crawl over my foot and I looked down and it was a cockroach," she recalls. "I opened my eyes and what was born was not me . . . [This new person] rose, she walked, she apparently talked. She was delighted. It is so ecstatic to be born . . ."

As she opened her eyes, Katie experienced the awe of rebirth. "I call it love," she says, "because I don't have another word. But just to see my hand in front of my face, or my foot, or the table, or anything, it's to see it for the first time. It's a privilege beyond what can be told. It's self experiencing the mere image of itself . . . born in love."

From that moment on, her suffering ended. She walked out of the halfway house a changed person. As she sums it up, "I discovered that when I believed my thoughts, I suffered, but that when I didn't believe them, I didn't suffer, and that this is true for every human being. Freedom is as simple as that. I found that suffering is optional.

I found a joy within me that has never disappeared, not for a single moment. That joy is in everyone, always."

Katie went on to become an international speaker and authored *Loving What Is: Four Questions That Can Change Your Life.* It describes "The Work," the self-inquiry method she developed as a result of her transcendent experience.

RACING YOUR CAR'S ENGINE IN THE PARKING LOT

For decades, one of the mysteries of neuroscience has been the energy consumption rate of the brain. Though it represents only 2% of our body's mass, it consumes 20% of its energy. It consumes *the same amount of energy* whether engaged in a demanding task or at rest.

An attention-intensive activity like updating your resume, counting your breaths, standing on one leg, composing a poem, or navigating a strange city at night requires the brain to rev up to maximum power. A set of brain regions called the Task-Positive Network or TPN gears up to conduct the task. It includes structures like the parietal lobe and the lateral prefrontal cortex. These handle functions like working memory, cognitive awareness, and attention.

But when you're sitting in your garden in a lounge chair on a Sunday afternoon, reading the comics or staring at the flowers in the garden, with your TPN dormant, your brain is *still* consuming 20% of your body's energy. MRI research shows that the brain's energy usage rarely varies more than 5% up or down in the course of a day.

It's like a car engine that keeps revving even when it's in idle. The Honda that Christine and I drove out of the fire has a six-cylinder engine with an output of 271 horsepower. When we're tearing down the expressway at 80 miles an hour, the tachometer tells us we're running at 2,000 revolutions per minute (RPM). But when we pull into the grocery store parking lot, put the transmission in park, and the engine is at idle, the engine is only turning over at 150 RPM, less than 10% of its peak output.

Your brain doesn't work that way. It's doing 2,000 RPM when you're in Park. Why? That has long been one of the puzzles of neuroscience.

It turns out that when you're doing nothing, your brain is not inert. It's highly active. But it's running on autopilot, with a set of regions called the

Default Mode Network (DMN) fired up. The DMN is what our brain *defaults to* when it's *not engaged in a task* like creating a spreadsheet or kicking a soccer ball or playing chess. When we're mentally at rest, our TPN shuts down while our DMN kicks in, keeping our brains running at 2,000 RPM even though we're not doing any specific mental task.

Marcus Raichle and the Dark Energy of the Brain

In the mid-1990s in a laboratory at Washington University in St. Louis, Missouri, neurologist Marcus Raichle made an unanticipated discovery. It wasn't the focus of his research, but suddenly, there it was. At the time, he didn't know what to make of it and merely filed away the lab results under a label he made up on the spot: medial mystery parietal area, or MMPA for short.

His accidental discovery led to him winning the Kavli Prize in neuroscience in 2014. This prestigious Norwegian prize is awarded for "outstanding scientific work" in the fields of astrophysics, nanoscience, and neuroscience.

What was the mystery discovery? Raichle was conducting PET (positron emission tomography) scans on the brains of subjects engaged in concentrated mental activities and on those of a control group at rest. To his puzzlement, the brain scans of the control group showed *increased* activity in certain areas when doing nothing. These areas were not active in subjects engaged in demanding mental tasks.

In describing this discovery, Raichle notes, "At some point in our work . . . I began to look at the resting state scans minus the task scans. What immediately caught my attention was the fact that regardless of the task under investigation, activity decreases were clearly present and almost always included the posterior cingulate and the adjacent precuneus."

The posterior cingulate cortex (PCC), an area around the midline of the brain, is linked to the limbic system, which is integral to memory and emotional processing. The precuneus is part of the parietal lobe and is also involved in memory. Raichle observed these shutting down during concentrated mental tasks and activating in a brain doing nothing in particular.

In 1997, Raichle and colleagues analyzed nine older studies and termed the collective regions of the brain that activated during a resting state and deactivated during attention-demanding tasks "the default mode network." Raichle also calls it the "dark energy" of the brain, referencing the astrophysics term for the mysterious, unseen force that comprises two thirds of the energy in the universe. The background energy or "dark energy" is what's consumed by DMN.

"It hadn't occurred to anyone that the brain is actually just as busy when we relax as when we focus on difficult tasks," observes Raichle. "When we relax, however, the DMN is the most active area of the brain."

It jumps around between thoughts, emotions, images, and memories. Meditators call it "monkey mind" after the way monkeys, never still, leap from tree to tree in the jungle. Why does the brain work this way? We don't yet know, though Raichle theorizes that such activity may help the brain stay organized.

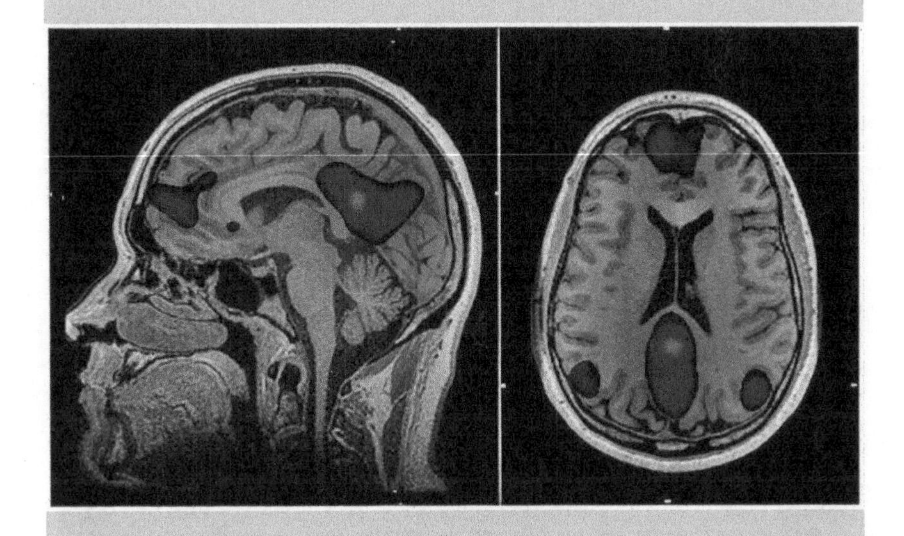

2.4. Brain regions active in the Default Mode Network.
See full-color version in center of book.

THE DEFAULT MODE NETWORK

The Emory researchers who identified the four phases of meditation found that when meditators slip out of the focused attention of the TPN and into Mind Wandering, the DMN activates.

The wandering mind of the DMN has a "me" orientation, focusing on the self. It may flit from what's going on at the moment ("Is that a mosquito buzzing?") to future worries ("I'm nervous about next week's exam") to the past ("I'm so mad at my brother Jim for calling me a sissy at my fifth birthday party").

The precuneus contributes to both self-referential focus and episodic memory. Disturbing memories are played and replayed. The idle brain defaults to what is bothering us, both recent and long-past events. These egocentric musings of the wandering mind form the fabric of our sense of self.

When you quiet your TPN in meditation, you open up a big empty space in consciousness. For a few moments, the brain is quiet, and you feel inner peace. Then the engine starts revving. The DMN kicks in, bringing with it a cascade of worries and random thoughts. You're doing 2,000 RPM in Park, but going nowhere.

And it gets worse. The DMN has a rich neural network connecting it with other brain regions. Through this, it busily starts recruiting other brain regions to go along with its whining self-absorption. It commandeers the brain's CEO, the prefrontal cortex. This impairs executive functions like memory, attention, flexibility, inhibition, planning, and problem-solving.

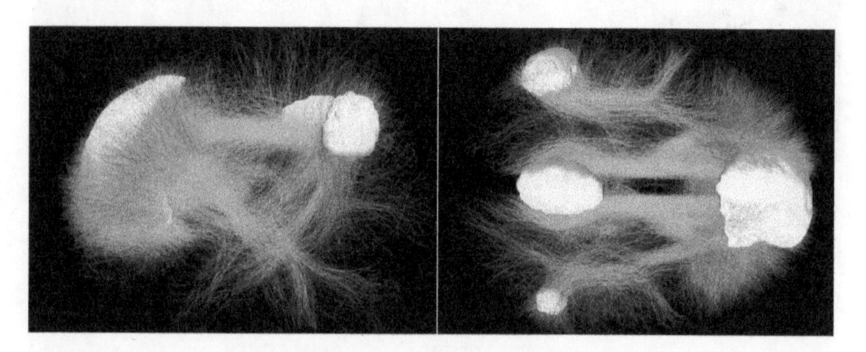

2.5. Nerves from the Default Mode Network reach out to communicate with many other parts of the brain. See full-color version in center of book.

The DMN also recruits the insula, a region that integrates information from other parts of the brain. It has special neurons triggered by emotions that we feel toward other people, such as resentment, embarrassment, lust, and contempt. We don't just *think* negative thoughts; we *feel them emotionally* too.

At this stage, the meditator isn't just wallowing in a whirlwind of self-centered thoughts. The DMN has taken the brain's CEO hostage, while through the insula it starts replaying all the slights, insults, and disappointments we've experienced in our relationships. The quiet meditative space we experienced just a few moments before has been destroyed.

This drives meditators absolutely nuts. No sooner do they achieve nirvana, the still, quiet place of Bliss Brain, than the DMN serves up a smorgasbord of self-absorbed fantasies. It pulls us into negative emotional states—then drags the rest of the brain along behind it.

The DMN. Hmm . . . that acronym reminds me of something: "the DeMoN." The DMN is the demon that robs me of the inner peace I'm seeking through meditation.

The Prince and the Demon

Buddhism has a story about prince Siddhartha, the future Buddha, on the eve of his enlightenment. When Siddhartha began his meditation practice that evening, a demon came to distract him.

This was no ordinary demon; it was Devaputra Mara, king of the demons. "Mara" is the Sanskrit word for "demon." It refers to anything that obstructs the attainment of enlightenment.

2.6. The demon attempts to shake the single-minded focus of the prince. See full-color version in center of book.

Mara knew that if he could disturb Siddhartha's concentration, he would pull him off the path to enlightenment.

Mara brought his whole gang, and they threw themselves enthusiastically into the task of knocking Siddhartha out of his single-minded focus. Some demons pitched arrows at him. Others shot fire. Yet others threw boulders, and when that failed to distract the prince, they picked up entire mountains and hurled them at him too.

Siddhartha's concentration remained on love. It turned the missiles into a rain of flowers.

Mara then tried a different tack. He took up the role of ringmaster, and ran the ultimate Miss Universe contest. He conjured up a wide variety of beautiful women. They were of all shapes, sizes, and colors. Black, yellow, red, white, short, tall, round, skinny—Mara offered Siddhartha any possible combination of pleasures.

The degree of Siddhartha's focus grew even deeper. He continued to meditate till dawn. In this state of perfect concentration, he forever removed the veil separating his local mind from nonlocal mind, and became a perfectly enlightened being.

Meditation traditions have evolved all kinds of techniques in an attempt to defeat the demon. Following the breath. Focusing on the third eye. Contemplating a sacred object. Repeating a mantra. Chanting. Walking slowly. Mindfulness. Invoking the name of a saint. All these are efforts to discipline the wandering mind.

Many of these techniques have a neurological effect. They *activate parts of the prefrontal cortex that govern attention*. This pulls the mind back to its meditative state. Research has shown that this regulatory neural connection is stronger in adept meditators than it is in novices. By refocusing your mind during meditation, you strengthen the neurological connection that presses the "stop" button on the demon's favorite movie, the Me Show. Even 3 days of mindfulness practice can increase this wiring.

While the demon is all about the self, meditation, especially loving-kindness meditation, is all about selflessness. All the mystical traditions of the world teach *nonattachment to self* as key to enlightenment. An area of the brain called the nucleus accumbens is part of the reward circuit regulating pleasure, addiction, and emotional attachment. Research shows that this area shrinks in longtime meditators. As they disentangle their attention circuits from the demon's endless "I, Me, Mine" attachment, they move into the transcendent state of Bliss Brain.

THE ME SHOW

Christianity has a story similar in theme to this one from Buddhism. On the verge of his maturity into public ministry, Jesus was tempted by the devil at the end of a solitary 40-day fast. He was hungry, and the devil suggested he turn the nearby stones to bread. Then the devil showed him "all the kingdoms of this world" and said they could all be his if Jesus would but bow down and worship him.

These wisdom stories from the world's great religions are more than ancient fables, or quaint superstition. They contain archetypal wisdom that illuminates core elements of the journey to enlightenment.

Notice the types of temptations the masters faced. The first attack by the devil played on Jesus's hunger. Mara presented the Buddha with his fears—everything that is going wrong. "The slings and arrows of outrageous fortune," as Shakespeare put it. That's the DMN's specialty: dredging up everything that has gone wrong in your past or might go wrong in your future. That's the first way the demon tries to tempt you out of Bliss Brain.

Then the demon presented Buddha with every possible variant of sexual and sensual pleasure. The devil offered Jesus all the wonders of the world. That's another way the demon tries to distract us out of focus. All the good things we might experience. If presenting you with all your fears fails, then presenting you with all your desires might succeed.

There's a final way the demon can yank us out of single-minded attention to focus. The brains of meditating monks show enormous amplitudes of gamma brain waves, about which we'll learn more in Chapter 4.

Gamma is the wave of insight and integration. In Bliss Brain, we have flashes of unparalleled insight. It's a creative brainstorm. You get downloads of brilliant blog posts you could write, extraordinary art you could paint, scientific breakthroughs you could achieve, marketing magic you might create, and life circumstances you might enjoy.

Yet going down these rabbit holes can be as much of a distraction as your fears and desires. It's all about me. My safety, my pleasure, my body, my money, my health, my love life, my career. Of all the streaming video series our minds could tune in to, the Me Show is the most compelling. It's the demon's ultimate weapon of mass distraction.

To reach and sustain Bliss Brain, it's essential to do what the Buddha and Jesus did: remain in one-pointed focus.

A WANDERING MIND IS AN UNHAPPY MIND

To track people's states of mind during their waking hours, two Harvard psychologists developed a smartphone app that contacted volunteers at random intervals. It asked them how happy they were, what they were doing at the time, and what they were thinking about. They could select from a menu of 22 common activities, like shopping, watching TV, walking, or having a meal. Eventually the researchers gathered some 250,000 pieces of information from 2,250 people.

What they found was that people spent about 47% of their time in negativity. The researchers wrote, "A human mind is a wandering mind, and a wandering mind is an unhappy mind."

When their TPNs were not focused on a task, their DMNs were focused on the self, immersed in their personal stories rather than the present moment. Their brains were working hard—but enchanted by the demon's unhappy stories. "The ability to think about what is not happening is a cognitive achievement that comes at an emotional cost," the researchers observed, concluding that "mind-wandering is an excellent predictor of people's happiness" (or lack thereof).

Mind Wandering takes you out of the present, and into regrets about the past or worries about the future. The constant chatter of Monkey Mind takes our attention away from whatever activity we're engaged in—and out of joy.

Participants in the study were least happy when they were doing one of three things. The first was tapping away on their personal computers, and the second one was work. No surprises there.

But the third one was surprising: resting. Why were they most unhappy when at rest? Without tasks to occupy their minds, the demon took over, recycling its endless unhappy absorption in self. When they escaped the demon's fixation on past and future, they came into the present moment. That's when they were at their happiest. Happy present moment activities included exercising, making love, and chatting with a friend. Sex was the only activity in which the demon consumed less than 30% of their attention.

2.7. The mind defaults to wandering between miserable past memories and fearful future possibilities.

The activity they were doing made little difference to their level of happiness when compared with whether or not their minds were wandering. Mind Wandering was more than twice as important in determining their degree of happiness than the activity. *Only in the present* could people find happiness.

WHY IS HAPPINESS SO FRAGILE?

If Bliss Brain is so desirable and pleasurable, *why is it so fragile?* Why can our brains be distracted from happiness by the slightest hint of a thought? Why is the demon's slightest whisper enough to drag us out of bliss? Why are our brains hardwired for negativity?

The answer is simple: That's how our ancestors survived. Those who were the most responsive to danger lived. If your ancestor's brain had a genetic mutation that heard the rustle of the tiger in the grass a nanosecond earlier, he started running a moment sooner. Genes that paid close attention to threats conferred an enormous survival advantage, as I illustrate in my book *The Genie in Your Genes.*

People who were less responsive to potential threats died, and their genes were lost to the gene pool. Those who reacted to the smallest hint of danger survived, passing their paranoid genes to the next generation.

In contrast, happiness provided little or no survival value. Fail to notice a beautiful sunset, ignore the sound of children singing, walk by a rose bush without smelling the blooms? Nothing bad happens.

But miss the rustle of the tiger? That's fatal.

So thousands of generations of evolution have honed our ability to respond to even the most minuscule whisper of the remotest possibility of threat, and abandon happiness at the drop of a hat. Mother Nature cares greatly about your survival—and not at all about your happiness.

That's why the DMN defaults to worry, instead of to bliss. Mentally rehearsing *future stuff that might just possibly hurt us*, past stuff that *definitely hurt us*, and present stuff that *might signal danger*—all these are signs of a brain that is successfully practicing the strategies that ensured our ancestors' survival.

This isn't bad. It's just excessive for the safe modern world in which we live. If you're at a construction site where a skyscraper is being built, you wear a hard hat and safety goggles. Such an outfit is entirely appropriate for that context. As attire for tea with the queen? Not so much.

Although the DMN interrupts meditation, it plays a useful role in our lives. It is active when we are thinking about others, considering our safety, remembering the past, and planning for the future. It is also active in self-oriented and social tasks, including memorizing the experiences we collect during task-oriented activities.

The path of your inner mystic will *elevate you to enlightenment*. The goal of your inner demon is to *keep you safe*. You can't get enlightened if you've been eaten by the tiger. Huxley observed that, "each one of us is potentially Mind at Large. But in so far as we are animals, our business is at all costs to survive."

People imagine they can modify the brain's negativity bias by positive action, but it's rarely successful. When you spend a weekend at a meditation retreat, read an inspirational book, or take an online happiness course, you can shift your state temporarily. But not for long. You're swimming against the tide of 4 billion years of evolution.

SELFING

That I-me-mine self is constructed largely in and by the brain's medial prefrontal cortex. It's assisted by the medial temporal lobe, the parietal lobe, and the PCC of which we'll hear more in Chapter 3. This brain network allows us to do things that other animals cannot. We can compose music and calculate math. We have a sense of time that includes past and future, allowing us to delay gratification to meet our goals. We are able to contemplate the very nature of consciousness, using the brain to think about our thoughts.

Yet consciousness is always turned on. Whether we're focusing on a task using the TPN or listening to the rambling of the demon, the engine is running at 2,000 RPM. There's no easy way of shutting off our thoughts, of getting outside the self.

In his book *The Curse of Self,* psychologist Mark Leary of Duke University shows the many downsides of this perpetual self-awareness. He shows that it leads to many forms of suffering, including "depression, anxiety, anger, jealousy, and other negative emotions." He concludes that self-awareness is "single-handedly responsible for many, if not most of the problems that human beings face as individuals and as a species."

We can summarize this state in a single word: "selfing."

Meditation quiets self-awareness and gives us relief from selfing. In experienced meditators, the "self" parts of the prefrontal cortex go offline. The jargon for this is "hypofrontality." *Hypo* is the opposite of *hyper,* and *hypofrontality* means the shutting down of the brain's frontal lobes. The inner critic shuts up. The negative self-talk about "who I am" and "what I do" and "what other people think of me" ceases. We quit selfing.

This gives us a sense of identity beyond the suffering self and all the roles it plays. Psychologist Robert Kegan is the former head of adult psychology at Harvard University. He calls the transcendence of selfing the "subject-object shift." In altered states, we get out of the subjective selves we normally think we are.

To be objective, you can't be the object you're contemplating. So when the brain enters a state of hypofrontality and we're no longer enmeshed in the local self, we gain perspective on it. We realize we're more than that. To realize it's an object we're observing, we have to step out of the suffering self. We see the demon from a distance as we step into an identity that is vastly greater than the one we previously inhabited.

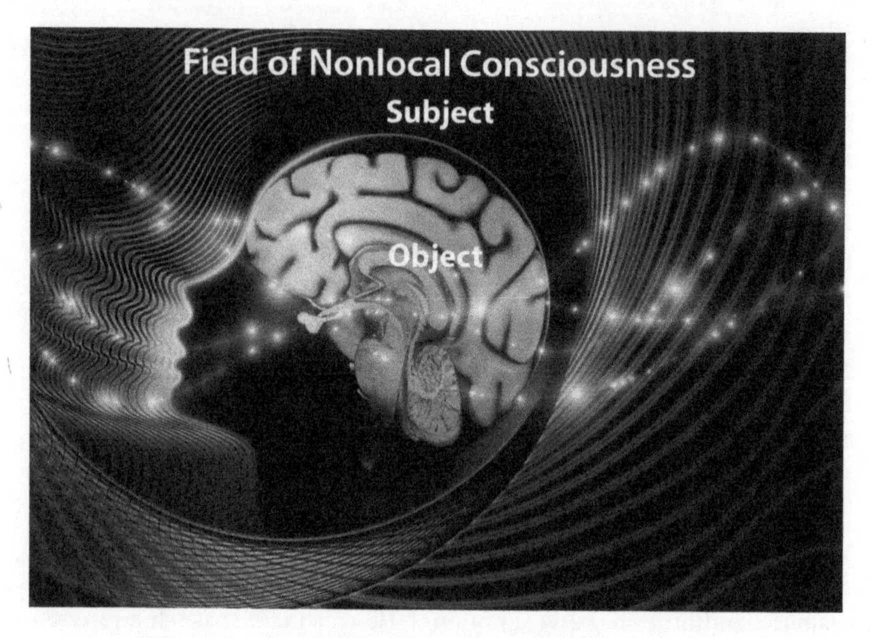

2.8. When we make the subject-object shift we escape the limitations of the finite self.

Kegan believes that making this jump is the most powerful way to facilitate personal transformation. He says that after it makes the subject-object shift, "the self is more about movement through different states of consciousness than about defending and identifying with any one form." This ability to let go of the conditioned thinking that keeps us trapped in the limited everyday self opens the door to an expansive nonlocal sense of self that encompasses our full potential.

Some people even experience leaving their bodies. Neuroscientist Andrew Newberg describes a Sufi mystic who suddenly found himself looking down on his body from the outside. Many meditators report "drifting upwards" or "traveling up the tunnel of light" or "seeing my body from a great height."

Newberg notes that when you make the subject-object shift, this "reduces activity in the fear-and-worry centers of your brain, and as you watch your anxiety, you become less anxious" and more able to transcend the suffering self. While writing about an AIDS study, I came across a similar story told by a patient.

No One Has a Monopoly on God

John was a gay, HIV-positive African-American man with a college education. One day, after transcending his preoccupation with his own suffering in order to help a drunk white man in distress, John had an out-of-body experience. Here's how he describes it:

"I felt like I was floating over my body, and I'll never forget this, as I was floating over my body, I looked down, it was like this shriveled up prune, nothing but a prune, like an old dried skin. And my soul, my spirit was hovering over my body. Everything was so separated. I was just feeling like I was in different dimensions, I felt it in my body like a gush of wind blows.

"I remember saying to God, 'God! I can't die now, because I haven't fulfilled my purpose,' and, just as I said that, the spirit and the body, became one, it all collided, and I could feel this gush of wind and I was a whole person again.

"That was really a groundbreaking experience. Before becoming HIV-positive my faith was so fear based. I always wanted to feel I belonged somewhere, that I fit in, or that I was loved. What helped me to overcome the fear of God and the fear of change was that I realized that no one had a monopoly on God. I was able to begin to replace a lot of destructive behavior with a sort of spiritual desire. I think also what changed was my desire to get close to God, to love myself, and to really embrace unconditional love."

This all happened in a moment, once John made the subject-object shift. Experiences of transcendence that allow us to see ourselves as distinct from the suffering self can completely reorder the fabric of our lives, changing behavior and personality. In Newberg's words, "self-reflective observation . . . activates structures in the brain directly associated with Enlightenment and transformation."

For centuries, Eastern religions have been telling us that it's our egos that trap us in suffering. In the 5th century, Indian adept Vasubandu wrote, "So long as you grasp at the self, you stay bound to the world of suffering." These spiritual traditions emphasize meditation, contemplation, altruistic service, and compassion as ways to escape the ego. Our emotions and thoughts become less "sticky" and "I, me, mine" "lose their self-hypnotic power." That's how we stop selfing.

Once we drop our identification with the ego-self enshrined in the prefrontal cortex and enter Bliss Brain, we make the subject-object shift. We can ask ourselves, "If I'm not my thoughts, and I'm the one thinking those thoughts, then who might I be?"

This perspective takes us out of selfing and into the present moment. In the meditative present, we can connect with the great nonlocal field of consciousness. Different traditions have different names for it: the Tao, the Anima Mundi, the Universal Mind, God, the All That Is. We then see our local self as the object.

With this view from the mountaintop, we're able to perceive new possibilities of what we might become, this time from the perspective of oneness with the universe. Free of the drag of the ego, uncoupled from the chatter of the demon, the conditioned personalities we inherited from our history and past experiences no longer confine our sense of self. Like John, making the subject-object shift allows us to rewrite our life script. It also makes us resilient in the face of life's upsets, as we'll discover in Chapter 8.

GETTING OUT OF YOUR HEAD

Given the torment the demon produces for us, it's not surprising that we want to get "out of our heads." Seven decades ago Aldous Huxley wrote that "Most men and women lead lives at the worst so painful, at the best so monotonous, poor and limited that the urge to escape, the longing to transcend themselves if only for a few moments, is and has always been one of the principal appetites of the soul."

Besides meditation, humans have devised a great many other escape routes. Among them are drugs and alcohol. According to a study by the RAND corporation, Americans spend close to $150 billion each year on illicit drugs. These include cocaine, heroin, cannabis, and methamphetamine. Another $158 billion is spent on alcohol.

In the book *Stealing Fire,* authors Steven Kotler and Jamie Wheal describe many other methods people use to tune out the incessant negative chatter of the DMN and break the spell of the local self. These include:

- Extreme sports like wingsuiting and kiteboarding
- Mood-altering prescription drugs like OxyContin and Adderall

- Therapy and self-help programs delivered both in person and online
- Gambling
- Video games
- Immersive visual experiences like pornography, 3D, and IMAX movies
- Social media
- Sex
- Group experiences like shamanic ceremonies, church revivals, and Burning Man
- States of heightened creativity in the arts and sciences
- Mystical experiences
- Communal performance such as dance, drumming, music, and theater
- Immersive art
- Neurofeedback and biofeedback

While from the outside these look like completely different experiences, from the inside they have much in common. They all produce positive changes in our internal felt experience. They put us in the present moment. They deactivate the demon and get us out of our heads.

Whether it's a Burning Man participant juggling fire, a monk meditating in a Himalayan cave, a wingsuiter flying through the Grand Canyon, a psychotherapy client breaking through to insight, an audience applauding the Three Tenors at the Metropolitan Opera, a computer coder seeing a vision of ones and zeros streaming through the Matrix after a caffeine-filled all-nighter, a channeler hearing the voice of an angel, a teenager raving to the beat of techno music, a nuclear physicist flashing on an elegant new theory, an OxyContin addict ascending a high, or a worshipper having a vision of Mother Mary at Lourdes, these experiences of getting out of our heads have a common neurobiological profile. They quiet the demon and give us a break from our endlessly spiraling thoughts.

Kotler and Wheal have long studied the inner state of flow through their Flow Genome Project. They are experts on peak performance and advisors to many top-level companies and organizations. Their research focuses on how we can escape the tyranny of the demon and incubate that delicious state of flow.

To measure how desperately we want to escape the chatter inside our skulls, they calculate how many dollars we spend on attempts to get out of our ordinary selves. They call this sum the Altered States Economy.

Their astonishing conclusion is that the Altered States Economy is worth $4 trillion per year in the United States alone; more than we spend on K–12 education, maternity care, and humanitarian aid—combined. That sum is greater than the size of the economy of India, or Russia, or the United Kingdom.

UCLA pharmacologist Ronald Siegel calls the need to intoxicate ourselves out of our ordinary states of consciousness the "fourth drive" of humans and animals, and it can be as compelling as our three other drives: for food, water, and sex.

Meditation is just one of the techniques people use to escape the demon, but it has qualities that make it uniquely effective. It's basically free; you can learn it online (see the resources at the end of this chapter), and it carries none of the financial penalties of gambling or consumables.

Meditation has no negative social consequences, unlike alcohol or drug use. Rather than the health penalties associated with addictive substances, it has positive side effects. It carries none of the risks of extreme sports. It's portable and can be used flexibly at any time or place desired by the user.

THE FOUR CHARACTERISTICS OF EXSTASIS

Studies of Tibetan monks with tens of thousands of hours of practice show that they're able to activate what Newberg calls the "Enlightenment Circuit," and take the selfing network offline. They deactivate the brain's PCC and medial prefrontal cortex, the neurological axis of the demon.

In this altered state, the parts of the brain associated with happiness, compassion, and equanimity light up. Kotler and Wheal describe four experiential characteristics of these ecstatic states. They are:

- Selflessness
- Timelessness

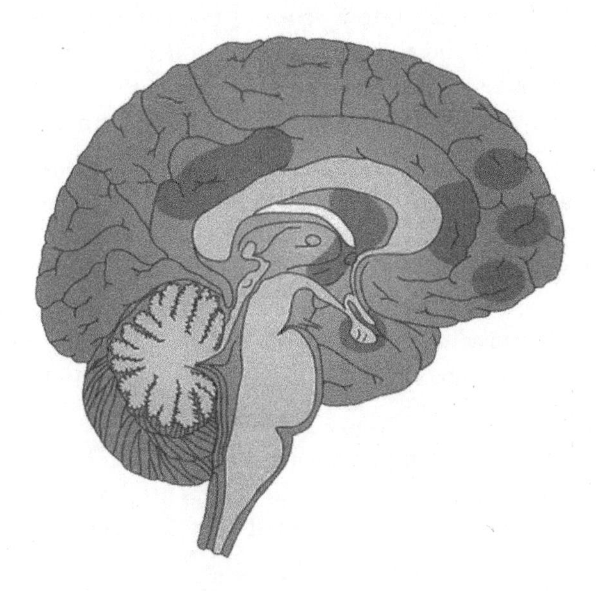

2.9. The Enlightenment Circuit associated with Bliss Brain. Brain regions include those involved with attention (insula and anterior cingulate cortex), regulating stress and the DMN (ventromedial prefrontal cortex and limbic system), empathy (temporoparietal junction, anterior cingulate cortex, and insula), and regulating self-awareness (precuneus and medial prefrontal cortex).

- Effortlessness
- Richness

They summarize these four qualities with the acronym STER. The benefit of this characterization of altered states is that it's not linked to a philosophy, religion, guru, or cult. It focuses on the *experiences* common to transcendent states, rather than *the paths by which people reach them*.

Selflessness represents a letting go of the sense of I-me-mine and all the elements that keep us stuck in our suffering default local personalities.

Timelessness means coming into the present moment. That's the place where we're free of the regrets of the past as well as worries about the future. We're in the timeless now, the only place we can experience the state of flow. In Huxley's words, "the eye recovers some of the perceptual innocence of childhood," while "Interest in space is diminished and interest in time falls almost to zero."

In this place, we relax into a sense of **effortlessness**. We feel connected to the universe and all living beings, our lives infused with a sense of **richness**. In this state we make connections between ideas, and the coordination between all the parts of our brains is enhanced.

These rich experiences feel deeply significant. Kotler and Wheal document the human drive for ecstasy as far back in time as the ancient Greeks, saying that Plato describes it as "an altered state where our normal waking consciousness vanishes completely, replaced by an intense euphoria and a powerful connection to a greater intelligence." Our English word "ecstasy" comes from the Greek words *ex* and *stasis*. It means getting outside (*ex*) the static place where your consciousness usually stands (*stasis*).

That's Bliss Brain. When you quiet the demon, you open up space in consciousness for connection with the universe. This produces a rich experience in which time, space, and effort fall away, and you merge with the rich infinity of nonlocal mind.

WINNING THE EXPERIENCE LOTTERY

One of the magazines I see at airports is called the *Robb Report*. It's a magazine for the ultrarich. It features glossy ads for $75-million yachts, private jets, designer clothes, and exclusive jewelry. You're still selfing, but now you're doing it in the presidential penthouse suite.

When I flip through the magazine, I reflect on how many children in developing nations could get clean water for the cost of one of those yachts. How many veterans could get the six sessions of EFT that it takes to cure PTSD? How many women could get shelter from domestic violence? How many near-extinct white rhinos could be protected from poachers? How many tons of plastic could be removed from the oceans?

Why do we want the yachts and jets? Why do we want that house on the hill or the flashy new car? Why do we want the glamorous girlfriend or the Rolex watch?

It's because we believe they will make us happy.

Dancing on the Moonlit Beach

On the way to teaching a workshop at Esalen Institute, Christine and I recently stayed with some friends in Carmel, California, a posh resort town by the ocean.

Shiloh McCloud is a visionary artist whose work in "intentional creativity" has inspired thousands of people to find their inner artist. Her husband, Jonathan, is a gourmet chef with an inquiring mind and a quick smile. After a heart-filled evening of conversation fueled by Jonathan's melt-in-your-mouth skirt steak and a bold malbec wine, the four of us enjoyed a sound night's sleep.

I woke up early in the morning before everyone else. I focused my mind on letting go of the demon and embracing Bliss Brain. As I entered flow, I was drawn to step outside the house. I smiled up at the moon and the huge glowing ring around it.

2.10. Dancing on the beach at dawn.

I listened to the crash of the waves on the cliffs below. The ocean beckoned. I walked down the path to the beach.

The demon kept reminding me of my problems. "You didn't tell anyone you were leaving the house," it reminded me. "What if you fall off the cliff in the dark? It could be hours before anyone finds your broken body on the rocks below. What if Christine wakes up and starts worrying?" I breathe and try to stay focused on the champagne air.

The demon doesn't quit. "You left your deodorant at home," it babbles. "Where will you get more on this desolate stretch of road between Carmel and Esalen? You'll be teaching all those students and you'll stink!"

On and on and on the demon drones, serving up an unending supply of doom and gloom. As soon as I release each problem, the demon presents me with another one.

When I get to the beach, I take off my sandals and enjoy the sand squishing through my toes. I fill my mind with gratitude, pushing out the demon's persistent cries. Spontaneously, my body flows into qigong postures. Awakening the Qi. Ringing the Temple Gong. Rolling the Steel.

The demon is forgotten. Fully in the present, I relax into the embrace of Bliss Brain.

I begin to sing, chanting a hymn of gratitude. A rogue wave crashes against a nearby rock. The backwash surges along the cliff face to soak me to the waist. I laugh at the kiss of the saltwater. I begin to run along the beach, dodging seaweed and rocks in the moonlight.

Out of breath, I rest. I call in my favorite archetypal guides and tune my mind to their frequencies. Joy. Vitality. Wisdom. Compassion. Beauty. Laughter. Integrity. Protection.

A half hour later, still in bliss, I walk back up the cliff path to my friends' house. The dawn is breaking, orange hues of cloud streaking the turquoise sky. Jonathan greets me at the door with a cup of steaming coffee. I walk into the bedroom and Christine opens her eyes. We stare at each other for several minutes, our eyes glowing with love. The day has begun.

Yet we can become happy every day without them. Happiness is an experience, and we can give it to ourselves at no cost. All we have to do to lose ourselves in Bliss Brain is quit selfing. We can train ourselves to orient to this state at the start of every day.

Such experiences are priceless. They make you happy in a way the yacht and the Rolex cannot. You can choose *exstasis* every day. You can start each day with gratitude and attunement to the universe no matter the distractions with which the demon tries to capture your mind.

Happiness is an equal-opportunity democracy. The bushman leading a subsistence Stone Age lifestyle in the Sahara can have it as surely as a Fortune 500 CEO. You can own the $75-million yacht and be much less happy than the dripping fool dancing on the moonlit beach for free. You can have nothing except soaked shorts and feel deliriously happy, immersed in Bliss Brain and free of the incessant chatter of the DMN.

It's *experience* that has value, not possessions. We desire possessions because we think they'll make us happier, but extensive research shows that once our basic survival needs are met, increased possessions don't boost happiness levels.

Meditation gives us the option of going straight to happiness and skipping the intermediate step of possessions. Acquiring them takes a lot of work and time, and all that effort can take us out of flow. We can spend a 40-year career amassing the possessions and money that we believe will give us happiness in retirement. Skipping the amassing stage and going straight to bliss gives us the end goal at the beginning. We win the gold medal before the contest even begins. Play doesn't happen in an imaginary future in which our lives are perfect. Play happens now.

We can become billionaires of happy experiences, the bank vaults of our minds overflowing with joy. That's the only currency that counts. We've then acquired the end state without going through the intermediate state of getting stuff. We've loaded the dice, so that any and every roll produces bliss.

Why not live like that every day?

DEEPENING PRACTICES

Here are practices you can do this week to integrate the information in this chapter into your life:

- **Releasing the Suffering Self:** That's the theme of this chapter's companion meditation. Use the link below to listen to this free 15-minute meditation each morning.
- **Play the "Name Your Demon" Game:** Give the selfing part of yourself a funny personal name, or ask it what its name is and write down the answer. One woman christened hers "Sticky." Another, "Yuggo." This exercise separates you from identification with the demon, and reminds you that you're in control.

- **Make the Subject-Object Shift:** Whenever you find your mind wandering during meditation, simply thank your DMN by name (e.g., "Thanks, Yuggo!") and then move your attention back to Focus.
- **Mindfulness App:** As a way of becoming mindful, enroll in the Harvard wandering mind study by using the link below to download the smartphone app.
- **Time in Nature:** Spend time in nature at least three times this week. Write those times in your calendar now, and treat them as seriously as you'd treat a doctor's appointment. This exercise in self-care is a way of centering your mind and nurturing yourself.
- **Journaling:** In your new personal journal, write down the insights you have this week. Notice the way your mind works in meditation, and describe it in your journal. Just a few words are enough, like, "Had a hard time getting to a good place this morning. Lots of mind wandering, but I settled down in 15 minutes."

EXTENDED PLAY RESOURCES

The Extended Play version of this chapter includes:

- Dawson Guided EcoMeditation: Releasing the Suffering Self
- Video of Qigong Routine
- Video of a Brief Introduction to the DMN
- Marcus Raichle on the DMN
- Full Version of Meditation Description at Start of Chapter. This includes two sections, "Communing with Archetypes" and "Seeking Answers from the Universe."

Get the extended play resources at: BlissBrainBook.com/2.

CHAPTER 3

ORDINARY ECSTASY

The human brain is an astonishing organ. It coordinates millions of physiological functions simultaneously every second. It's the interface between our consciousness and our bodies.

Your brain is 75% water. Of the remainder, 60% is fat. It represents 2% of your body weight yet consumes 20% of your energy. To feed its voracious energy needs, it has around 100,000 miles of blood vessels through which it absorbs 20% of the body's oxygen supply. It generates up to 23 watts of power, enough to light up a room.

Gray-colored neural tissue, or gray matter, makes up 40% of its volume. The other 60% is white matter.

When growing during early pregnancy, the human brain is adding 250,000 neurons a minute. Between birth and 18 years of age, it triples in size. Although it doesn't grow bigger after that, it keeps adding surface area in a series of ever-more-complex folds.

Neurons connect with each other using extensions called synapses. A single neuron can have as many as 7,000 synapses and network with thousands of other neurons in this way.

New synapses may be formed rapidly, and when you stimulate a neural bundle by passing information through it, the number of synapses increases. In an hour of repeat signaling, the number of synaptic connections can *double*. As you use neural pathways, they grow.

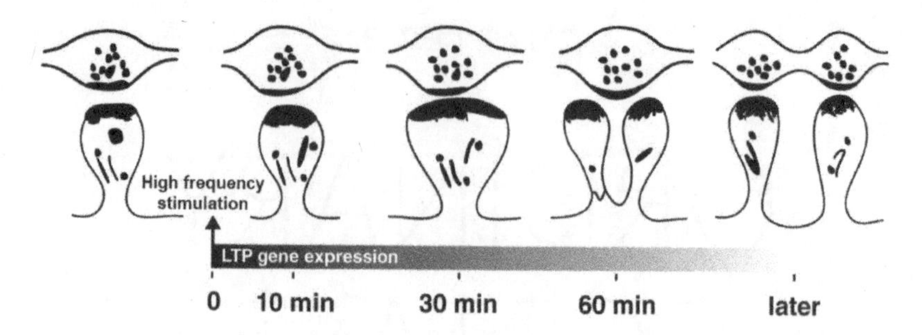

3.1. In an hour of repeat stimulation, the number of synapses in a neural bundle can double.

If you stop using a neural pathway, it begins to shrink. After about 3 weeks, the brain notices you're not sending information through that channel anymore and begins to disassemble the cells of which it is composed.

As a result, the *brain regions you use most grow larger and/or signal faster,* while the ones you don't use atrophy and shrink.

The brain and its associated nervous system began evolving some 600 million years ago. The first part to evolve was the hindbrain, including the brain stem and cerebellum. It coordinates survival functions such as respiration, digestion, reproduction, and sleep. The dinosaurs had hindbrains much like our own. So do their descendants, lizards and birds. That's why the hindbrain is often called the reptilian brain. It is so good at *regulating survival functions* that it hasn't changed much in millions of years.

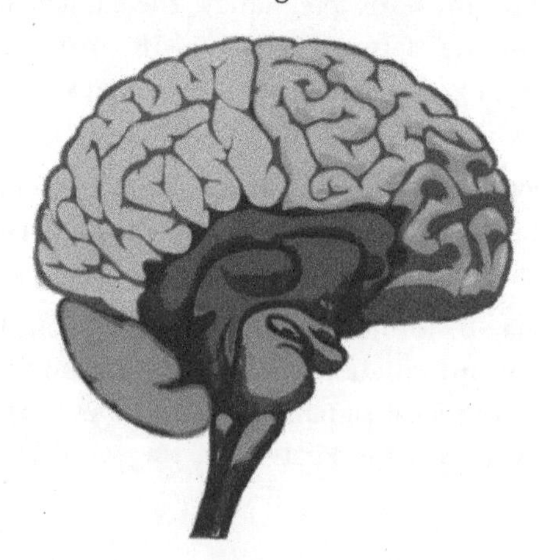

3.2. The three-part brain. See full-color version in center of book.

Stacked above the hindbrain is the midbrain or limbic system. It handles memory and learning, as well as emotion. Within the limbic system are several substructures with specialized functions, such as the thalamus, hypothalamus, hippocampus, and amygdala. It evolved more recently, about 150 million years ago, with the rise of mammals. Mammals can do things reptiles can't, like regulate their body heat, give birth to live young, form strong emotional bonds with each other, and learn from emotional experiences.

The biggest part of the human brain is the neocortex. It is also the newest, in terms of evolutionary development. The evolution of primates such as apes and monkeys led to a growth in brain size, which accelerated in our hominid ancestors. Since early humans began making tools about 3 million years ago, the brain has tripled in size.

The neocortex handles conscious thought, language, sensory perception, and generating instructions for the other brain regions. Cats have around 4% of the volume of this type of brain tissue as humans, while dogs have 7%. Apologies, cat people.

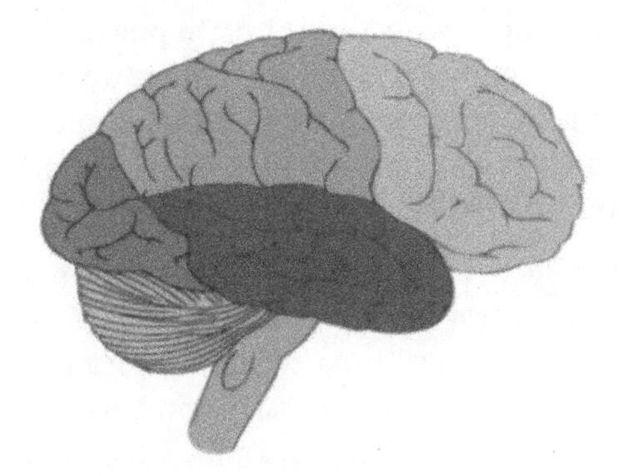

3.3. The four lobes of the cortex. See full-color version in center of book.

The brain's cortex is divided into four lobes: the frontal lobe (front of head), parietal lobe (top of head), occipital lobe (back of head), and temporal lobe (sides of head). Each one has distinct functions. The occipital lobe processes information from our eyes. The temporal lobe integrates memory with input from our senses of sound, touch, taste, smell, and sight. Information about movement, temperature, taste, and touch is processed by the parietal lobe. Executive functions like cognition and the control of our actions are the province of the frontal lobe, which is also the seat of what we think of as "self."

THIS IS YOUR BRAIN ON BLISS

During the past century, advanced brain imaging systems like functional magnetic resonance imaging (fMRI), electroencephalograph (EEG), positron emission tomography (PET), and magnetoencephalography (MEG) have been developed. These allow scientists to determine which brain regions are active and which ones are inactive during various states.

Researchers measure which parts of the brain are engaged during decision-making, sleep, stress, paying attention, daydreaming, insight, play, social interaction, and emotional experiences. They've also turned their attention to finding out how the brain works during meditation and other transcendent states.

Studies of spiritual adepts such as Tibetan Buddhist monks and Franciscan nuns provide data on the brain states common to spiritual practices. The stories of famous figures such as Albert Einstein and Babe Ruth also provide insights into how their prodigious abilities were related to enhanced function in particular parts of their brains.

During transcendent experiences, the brain processes information differently. This shows up on brain scanners as deactivation of some regions and activation of others. After examining the brain function of thousands of meditators, researchers now have a composite picture of how meditation changes the way the brain works.

Brain structures such as the amygdala, which turns on the fight-or-flight response, and the parts of the parietal lobe that process sensory input shut down. Others—such as the insula, which is associated with happiness, kindness, and compassion; the corpus callosum, which connects the left and right hemispheres; and the hippocampus, which handles emotional self-control—light up.

The parts of the prefrontal cortex (PFC) that keep your attention focused and regulate emotion are active during meditation, while the parts that synthesize your personality—selfing—go dark. In this chapter, we'll review each of the parts of what neuroscientist Andrew Newberg calls the brain's "enlightenment circuit" in turn.

Examining each region of the Enlightenment Circuit allows us to build up a comprehensive picture of how our brains function when we meditate. (In Chapter 6, you'll learn about the four networks comprising this circuit and how parts of the brain actually increase in volume with meditation.)

We then understand how we can meditate most effectively and trigger the extensive health and happiness benefits of meditation.

THE PARIETAL LOBE: BRIDGE TO THE OUTSIDE WORLD

The parietal lobe sits at the top of the head and plays an important role in helping us relate to the world outside. It is the place where information from our senses arrives and is integrated with information from our internal states.

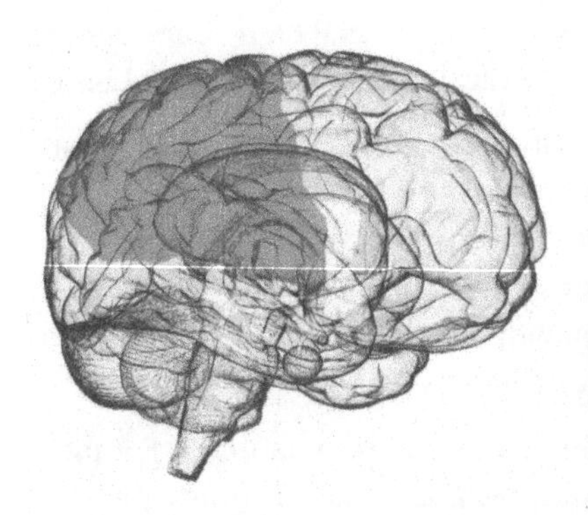

3.4. The parietal lobe: bridge to the outside world.

The parietal lobe receives input from our senses, perceiving touch, taste, and temperature, as well as determining where we are in relation to the world around us. When you reach out your hand to pick up a glass of water, the parietal lobe is determining where your fingers end and the glass begins. People with damage to their parietal lobe can be unable to perform simple functions like picking up objects.

When you dance with a partner, scratch a mosquito bite, use your GPS to navigate to an unfamiliar destination, step onto a curb, stick your fork into a meatball, or stroke your child's head, your parietal lobe guides the movement of your body in space. It gauges angles and distances. It is responsible for defining the boundary between self and other.

Your brain's parietal lobe coordinates the actions of your limbs with your eyes and also has a role in processing language and attention. Even without visual input, the parietal lobe can determine the location of familiar objects. You can easily test this yourself. Close your eyes and touch your cheek. You're able to move your hand in perfect coordination even without visual input.

In longtime meditators, such as Tibetan monks who have meditated for over 10,000 hours, the parietal lobe shows greatly reduced activity during meditation. While the brain's energy usage only fluctuates around 5% up or down each day, in these adepts it drops up to 40% as they enter an altered state of consciousness.

People who are isolated and lonely show the opposite effect. Their parietal lobes may be highly active. Feelings of loneliness and isolation are increasing in our fragmented society, according to a number of studies, and they have detrimental effects on our health. We are social beings, and a sense of connectedness translates into overall physical and emotional well-being.

In a meta-analysis of 148 studies with a total of 308,849 participants, the researchers found "a 50% increased likelihood of survival for participants with stronger social relationships. This finding remained consistent across age, sex, initial health status, cause of death, and follow up." Even when the researchers corrected for behaviors such as smoking, obesity, and lack of exercise, the effect remained consistent.

Think back to the times of your life when you were happiest. For most of us, special moments with family and friends come to mind. We may have been on vacation with loved ones or enjoying a meal or a joke with friends. It may have been singing carols during the holidays, when billions of people join in affirming "peace on earth, goodwill to men."

What's common to all those happy times is that your attention was fully in the present moment, the "timelessness" of Chapter 2. You weren't worrying about the past or stressing over the future, the way you might be doing if you didn't have precious people to engage your thoughts. Time and space receded as the love-filled present occupied your attention.

Only in the present moment can you escape the demon's obsession with the mistakes of the past and the problems of tomorrow. Experiences of timelessness, as the parietal lobe shuts down, shape our perception of the world and how we act in it. A Stanford study of people who experienced "the deep now" found that it changed their behavior. They "felt they had more time available . . . and were less impatient . . . more willing to volunteer their time to help others . . . preferred experiences over material products . . . and experienced a greater boost in life satisfaction."

A quiet parietal lobe promotes empathy, compassion, relaxation, appreciation, connectedness, and self-esteem.

Andrew Newberg, the neuroscientist who has studied both Tibetan monks and Franciscan nuns using advanced fMRI and SPECT (Single-Photon Emission Computed Tomography) scans, says that "when people lose their sense of self [in meditation], feeling a sense of oneness . . . this results in a blurring of the boundary between self and others . . . [with] no sense of space or passage of time."

Worshipers in meditation literally lose themselves in a sense of oneness with the universe, as shown by the quieting of their parietal lobes. The Franciscan nuns use the Latin of their faith to describe the experience, *unio mystica* (mystical union). Buddhists call it "unitary consciousness."

Newberg says that "Unity is . . . the foundational notion in pretty much every religious tradition . . . people often describe unity as more 'fundamentally real' than anything else they've ever experienced. More real than reality." In a survey of 2,000 people who had enlightenment experiences, over 90% report that they're more real than everyday reality.

This sense of hyperreality may occur because in these states the energy saved by the 40% deactivation of the parietal lobe is redeployed to enhance attention. Newberg says, "It's an efficiency exchange . . . energy normally used for drawing the boundary of self gets reallocated for attention. At that moment, as far as the brain can tell, you are one with everything." Newberg notes that this produces a sense of awe.

Meditators don't feel alone. They describe a sense of connection with everyone and everything in the universe, part of an infinitely large picture. On the material level, like everyone else, you still have vibrant connections with family and friends. But you also have a source of social support that isn't dependent on the availability of other human beings. As you merge with universal consciousness, you feel one with all beings and the universe itself.

An overview of meditation studies by a panel of top experts shows that this opens up "what have been termed 'nonlocal' aspects of human consciousness . . . people report experiences of perceiving information that does not appear limited to the typical five senses or seems to extend across space and time, such as precognition, clairvoyance, and mind-matter interactions (described as 'siddhis' in the Hindu yogic traditions)."

Losing My Personality and Finding Bliss

One of the most famous articulations of the experience of losing attachment to one's personality comes from brain scientist Jill Bolte Taylor. In 1996, a blood vessel burst in her brain, shutting down her left hemisphere.

As her parietal lobe went offline, the personality she'd embodied for the previous 37 years wasn't accessible anymore. She could not remember who she was or her life's history. She could not speak or understand language. Without her parietal lobe's ability to determine where she ended and the floor began, she could not walk.

Far from being terrified, she found the experience liberating. Her neuroscience background equipped her to tell her story from inside the experience. She described it in a TED talk, which went viral. She wrote a book titled *My Stroke of Insight*, which became a *New York Times* bestseller. She reflected on the event in an interview in *Unity* magazine:

"The consciousness of my right brain did not recognize the boundaries of my body at all. The left parietal region of the brain, in what's called the orientation association area, holds a holographic image of your body so that you know where you begin and where you end. When those cells went offline after the stroke, I no longer had that perspective.

"I felt as big as the universe! My body was attached to me, but I didn't experience it as my essence. Instead, I was the collective whole, connected to everyone and everything—I was completely fluid. Our right brain doesn't see the artificial division of individual bodies that the left brain places on us. We're actually all energy. Our bodies are just energy compacted into a dense form.

"During the stroke, I felt enormous, like a genie who had just come out of its tiny, little bottle. That's the power of what we really are . . . We are all this enormousness that gets squeezed inside these tiny little bodies and we think this little, tiny body is what we are. But it's really just the tool we use to do stuff in this physical world."

Of the 1,120 meditators surveyed in the overview, 56% reported experiences of clairvoyance or telepathy; 82% increased synchronicities; 86% an altered sense of time, and 89% the moments of insight typical of gamma brain waves.

More practitioners reported contact with nonphysical entities than with a real-life meditation teacher. Over half had multiple contacts with a being they characterized as a guide, an angel, God, or a higher power. The local mind of the meditator becomes one with the nonlocal mind of the universe.

This gives the meditator a profound sense of connection in both states; whether interacting with other human beings in everyday life or interacting with the universe in meditation. Taking your parietal lobe offline puts you in the present moment, by detaching your awareness from time and space. Meditators lose their sense of time in a feeling of timelessness. As your small local self merges with the nonlocal universe, your sense of space expands to the borders of infinity. Deactivating the parietal lobe anchors you in the present moment.

Parietal Lobe Deactivation Benefits:

- Feeling of oneness
- Centering in present moment
- Sense of connectedness
- Empathy
- Compassion
- Self-esteem
- Happiness

- Loneliness
- Social anxiety

THE CORPUS CALLOSUM: THE HEMISPHERIC LINK

The corpus callosum links the right and left hemispheres of your brain. It represents the largest aggregation of white matter in the entire brain. There are differences between the two hemispheres of your brain, and they communicate through the corpus callosum.

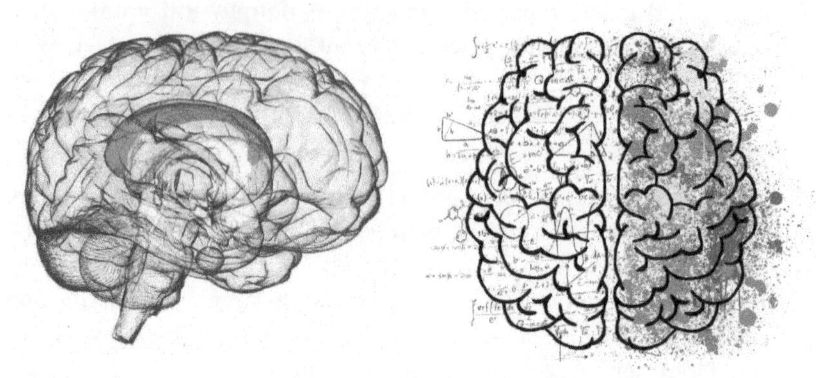

3.5. The corpus callosum: exchanging information between hemispheres.

This integration is important because some functions aren't evenly distributed between left and right hemispheres but are dominant in a single hemisphere. For instance, language and speech are generated primarily in the left hemisphere. The right hemisphere, on the other hand, specializes in nonverbal communication and emotion. A well-developed corpus callosum integrates information from the two hemispheres and balances their activity.

When we're meditating, our corpus callosum lights up. We're exchanging information between the right and left hemispheres of our brain. This allows us to make connections and balance the function of the two halves of the brain. A study comparing 30 meditators with 30 non-meditators found that the meditators had a significantly greater amount of tissue in the corpus callosum. Creative people are also better at integrating information from many sources to arrive at novel solutions to problems.

The corpus callosum contains a structure called the anterior cingulate cortex (ACC), which plays a role in our ability to feel empathy. The ACC is activated by the hormone oxytocin, often called "the bonding hormone" because contact with people we care about stimulates its release. The ACC is one of the structures lit up in the Enlightenment Circuit.

Greater corpus callosum volume is associated with intuition, creativity, focus, memory, problem-solving ability, and coordination. When you meditate, you activate the corpus callosum. This gives you access to all of the resources available in both hemispheres of your brain, as well as the ability to integrate the information you receive.

Babe Ruth's Brain

Babe Ruth, also known as the Sultan of Swat, played 22 Major League Baseball seasons, from 1914 through 1935. His all-time record for the most seasons with more than 40 home runs (11 of them) stands to this day. He is widely viewed as the greatest baseball player of all time.

In 1921, *Science* magazine reported on research into what made Babe Ruth's batting swing so phenomenal. Investigators Albert Johanson and Joseph Holmes of the psychology laboratory at Columbia University subjected Ruth to a battery of physical and psychological tests. They traced his abilities to these findings:

- Ruth's physical actions were 90% efficient compared with a human average of 60%.
- His eyes functioned approximately 12% faster than the eyes of the average person.
- His ears operated at least 10% faster than the ears of the average human.
- His nerves were steadier than the nerves of 499 out of 500 other people tested.
- He was one and a half times above the average in attention and quickness of perception.
- He was about 10% above normal in intelligence, as evidenced by the rapidity and accuracy of his understanding.

Babe Ruth's eyes and ears operated faster than the eyes and ears of other people. His brain recorded sensory input from his body more rapidly and sent the subsequent orders to his muscles faster. The coordination between Ruth's eyes, brain, nerves, and muscles was near perfect.

He was tested on a very warm night after the exhaustive play of an afternoon baseball game during which Ruth hit a home run. The testing, held in a stuffy room, went on for over 3 hours and required Ruth to stand for most of it, as well as walk up and down stairs five times.

Observers reported that Ruth exhibited no sign of tiredness and noted that the already impressive results would likely have been even better had he been fresh.

The tests also revealed that Ruth held his breath when he swung the bat. We now know that such constriction inhibits a swing's power. His already superhuman swat would have been even better if he hadn't held his breath.

Ruth was a left-handed batter and thrower. Left-handed as well as ambidextrous people have 11% more tissue in the corpus callosum than right-handed people.

Corpus Callosum Activation Benefits

- Focus
- Memory
- Creativity
- Communication between left and right hemispheres
- Intuition
- Perceptual clarity
- Happiness
- Musical ability
- Coordination
- Integration of information
- Ambidexterity
- Vision

- Dyslexia

Albert Einstein also had an enhanced corpus callosum. Researchers compared his brain with those of 15 men all 76 years old, the age at which Einstein died, as well as with 52 young, healthy men. His corpus callosum was a superhighway of connectivity, "thicker in the vast majority of subregions" than those of the other brains.

THE HIPPOCAMPUS: TRACKING THREATS AND MAKING MEMORIES

The hippocampus is a seahorse-shaped structure in the center of the limbic system. In my live workshops, the analogy I pick for the hippocampus is that of the military historian. Its most vital job is to *compare incoming information* with the *memory of past threats*. If there's a match, it sounds the alarm by activating the amygdala, which in turn switches on the whole fight-flight-freeze (FFF) system.

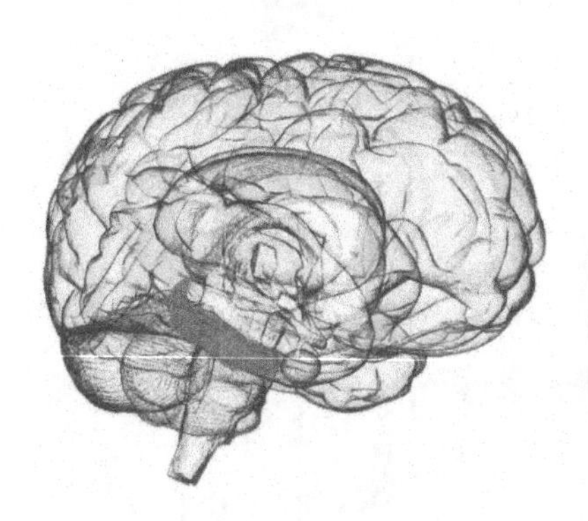

3.6. The hippocampus: tracking threats and recording memories.

By deciding which signals to pass to the amygdala and which to ignore, the hippocampus regulates our emotions. Some people have an active hippocampus that effectively regulates emotion. Others do not; these unfortunates have a hair-trigger response to their own emotions. They become angry, fearful, or anxious at the slightest stimulus. Their behavior is dictated by their emotions.

The hippocampus is also the seat of learning. Novel experiences produce the growth of new synaptic connections in the hippocampus. Go take a class in Mandarin Chinese, learn pickleball, date a new love interest, experiment with recipes from a Hungarian cookbook—your hippocampus will start to grow new connections.

But the most essential function of the hippocampus is to *catalog the bad stuff of the past,* and if anything coming our way in the present resembles that bad stuff, it *makes a match* and turns on the FFF response.

The Aircraft Identification Chart

Growing up in the post-WWII British Empire, I was surrounded by reminders of the war. Retired field guns mounted on either side of the entrance to government offices. Penny whistles made out of spent shell casings. Steel helmets in the closets of relatives. Airplane and tank models made from bullets.

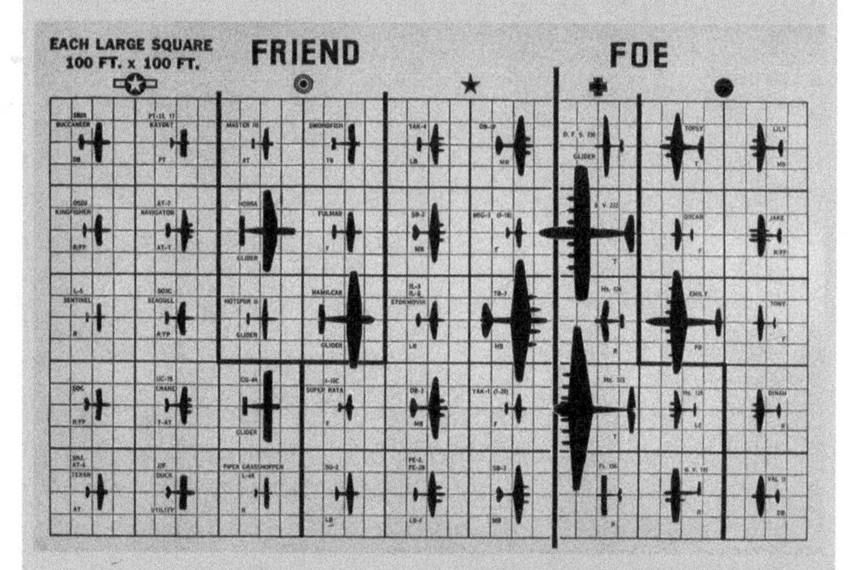

3.7. Aircraft Identification Chart.

One of these reminders was the aircraft identification chart. When Britain was being bombed during the Blitz, the sound of aircraft engines overhead was commonplace. When people looked up at the night sky, they might see the silhouette of a plane overhead. Knowing whether that shape was a British or a German aircraft could mean the difference between life and death.

The wartime government put aircraft identification posters everywhere, and even in the 1960s I remember seeing them frequently in forgotten corners of pubs or hotels. Before I was 2 years old, I could tell the British Lancaster bomber from the German Heinkel He 111 at a glance. When as a toddler I was taken to the Imperial War Museum, I could glance at any WWII aircraft and name it.

By that time, however, such knowledge was useless. The war had been over for 10 years and there was no survival value in memorizing the shapes of the aircraft on the charts. Yet they were engraved indelibly in the neural pathways of my hippocampus.

That type of pattern recognition is the job of the hippocampus. It compares incoming information, such as the shape of an airplane, with its huge encyclopedia of past knowledge. If it's a threat—"Heinkel 111!"—it sends a signal to the amygdala, which aborts every other physiological function as it screams, "Drop everything! Run for the air raid shelter!"

If there's no match with past threats—"Lancaster"—the hippocampus remains quiet.

When we learn about threats as children, and they are accompanied by strong emotions such as fear, they can remain embedded in the neural circuits of the hippocampus for life. Neuroscientists call these "deep emotional learnings."

Like the old posters, they may have no use in the present. They may even be triggering us to react to threats that are entirely imaginary. Yet once learned, and reinforced by conditioned behavior, they are hard to change. Like the dusty posters in the pubs, they may hang around long after they've outlived their usefulness.

When the hippocampus isn't sure what to make of a piece of information, it refers it to the brain's prefrontal cortex (PFC). That's the brain's executive center, the seat of discrimination and knowledge. It takes incoming information from the hippocampus and determines whether the apparent threat is real.

For instance, you hear a loud bang and are immediately alarmed. "Gunfire?" wonders the hippocampus. "No," the PFC tells it. "That was a car backfiring." The reassured hippocampus then does not pass the alarm to the amygdala. Or perhaps the PFC says, "That group of young men hanging out in the parking lot looks suspicious," and the hippocampus then signals the amygdala, which puts the body on Code Red.

Using that path from the emotional center of the brain to the executive center is crucial to regulating our emotions. Because it involves a feedback loop with information going first to the PFC and then back to the hippocampus from the PFC, it's called the *long path*: hippocampus > PFC

> hippocampus > amygdala > FFF. The long path is the default for people with effective emotional self-regulation.

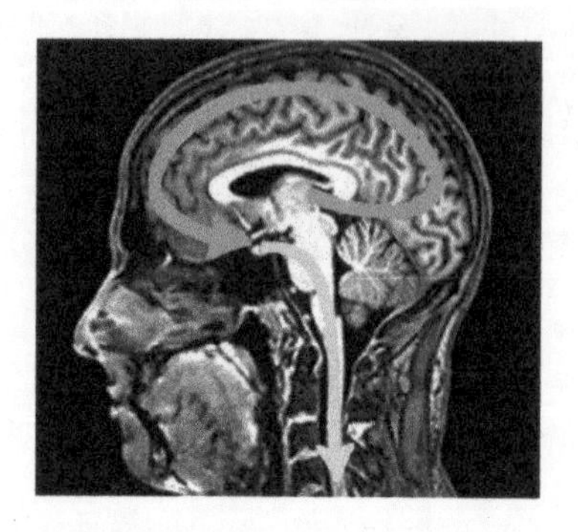

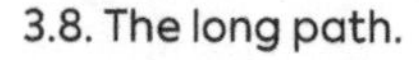

3.8. The long path.

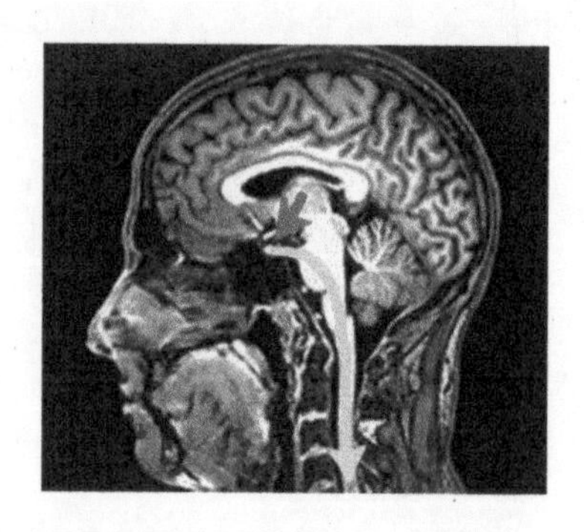

3.9. The short path. See full-color version in center of book.

In people with poor emotional self-regulation, such as patients with PTSD, this circuit is impaired. They startle easily and overreact to innocuous stimuli. The hippocampus cuts out the PFC. Instead of referring incoming threats to the wise discrimination of the primate brain, where the bang can be categorized as "car backfiring," the hippocampus *treats even mild stimuli as though they are life-threatening disasters* and activates the amygdala. This short-circuit of the long path creates a *short path*: hippocampus > amygdala > FFF. The short circuit improves reaction speed, but at the expense of accuracy.

The hippocampus is also central to memory and learning. It helps us process at least two different types of memory: declarative memories and spatial memories. Declarative memories have to do with events and facts such as memorizing your lines in the school play or the rankings of the teams in the World Cup.

Spatial memories have to do with navigation; they help cab drivers memorize routes through the heart of a medieval city or allow violinists to know exactly where on the strings to place their fingers.

That's evidence of another remarkable feat the hippocampus performs: turning short-term memories into long-term ones. The first few violin lessons, you have to focus intently to place your fingertips even close to the correct spots. With experience, once you've engraved the information in long-term memory, you can play effortlessly, without thinking. The hippocampus can control function in several distant parts of the brain simultaneously, coordinating their activities.

When we meditate, we activate the hippocampus, in particular a subregion called the *dentate gyrus.* It's function is to synchronize emotional regulation in different parts of the brain, and we'll discover just how amazing it can be in Chapter 6. This synchronization means we're able to calm our turbulent emotions, activate the long path, and give our consciousness a break from the love and fear, envy and desire, and resentment and attraction sweeping through the Default Mode Network (DMN). Without these emotions to distract us, we create a calm emotional space for Bliss Brain.

Hippocampus Activation Benefits

- Memory
- Learning
- Emotional control
- Skill acquisition
- Spatial navigation
- Sense of direction
- Serotonin
- Coordination of brain regions

- Depression
- Hyperactivity
- Amnesia
- Alzheimer's

THE AMYGDALA: THE FIRE ALARM

The amygdala is often referred to as the "fire alarm" of the brain. It is activated by signals that other areas of the brain, primarily the hippocampus, interpret as danger. If you're swimming in the Caribbean sea and see a large dorsal fin sticking out of the water, the hippocampus compares it to similar images in a flash. "Shark!" it concludes and sends a signal to your amygdala.

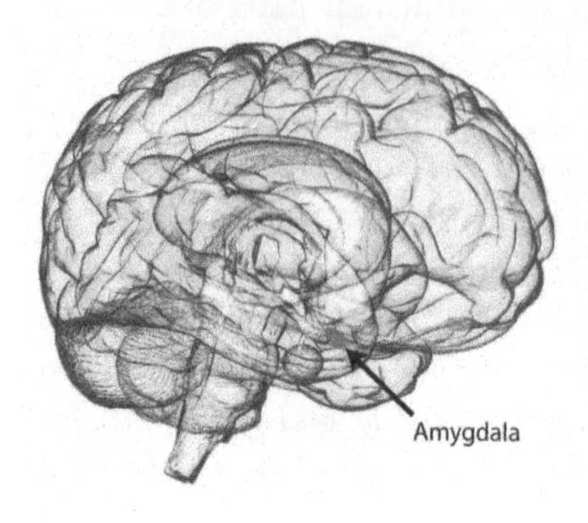

3.10. The amygdala: the fire alarm.

Nerves from the amygdala run to the parts of your brain stem that govern the fight-flight-freeze (FFF) response. From there they radiate out to every single organ system in your body: your heart, lungs, muscles, digestive tract, immune system, and skin.

When your body goes into FFF, every one of those organ systems is recruited, because survival is the paramount drive of every living being. Your lungs furiously pump oxygen into your bloodstream, aided by your heart, which speeds up to increase the blood supply to your muscles. They engorge with blood, especially the large muscles of your arms and legs, as they get ready to either run or fight.

Your digestion shuts down. The tiny capillary veins and arteries that carry blood to your gut shrink by as much as 80% as your body deploys precious resources away from nonessential systems such as digestion to essential ones such as circulation and respiration.

Your immune system shuts down too. One study found that couples who fight have a drop of up to 40% in their immune function. Most parts of your PFC shut down as capillaries there contract; your body has no use for the PFC's ability to compose a symphony or perform calculus when the first order of business is to escape the shark.

If you have good emotional self-regulation, and your long path is working, your hippocampus refers the sighting of the fin to the PFC, which says "dolphin." Once the hippocampus gets the information that the stimulus is innocuous, that the fin belongs to a playful dolphin rather than a predatory shark, it doesn't activate the amygdala and you remain calm.

If your long path is short-circuited by stress, and your brain is using the short path instead, you might be so alarmed at the mere thought of a shark that you have a panic attack just thinking about taking a swim in the ocean.

All the body's machinery of FFF then gets engaged by this imaginary threat, just as if you were nose to nose with Jaws. Your gut clenches, your heart races, your breathing becomes fast and shallow, and your focus narrows to the point where you can't think about anything other than the threat.

This takes a huge biological toll on the body. High adrenaline produces dramatic reductions in life span. Stressed people have much more disease and live much shorter lives than unstressed people. Whatever form stress takes—depression, anxiety, or PTSD—correlates with higher rates of cancer, diabetes, and heart disease. The deficits in the life spans of stressed people are measured in decades rather than years.

In meditators, the amygdala is quiet. It becomes even quieter with practice. The difference in amygdala activation between the longest-term meditators and their less-experienced peers has been measured. The adepts show 400% less reactivity to stressful events. But even in novices who practice mindfulness for 30 hours over 8 weeks, decreased amygdala activity is found.

Other structures within the midbrain or limbic system work together with the hippocampus and amygdala. One of them, the thalamus, is like a relay station. Close to the corpus callosum, it identifies information coming in from the senses like touch, hearing, and taste, and directs it to the consciousness centers of the prefrontal cortex. The thalamus typically becomes more active during meditation, as it works harder to suppress sensory input (like "that buzzing mosquito" or "this chair is too hard") that pulls us out of Bliss Brain.

With the hippocampus regulating emotion, the thalamus regulating sensory input, and the long path in good working order, stress-inducing signals aren't sent to the amygdala. In turn, all the body's FFF machinery remains offline. This produces corresponding biological benefits. Heart rhythm is even. Respiration is deep and slow. Digestion is effective. Immunity is high. That's why so many studies show pervasive health and longevity benefits among meditators.

Amygdala Deactivation Benefits

- Calmness
- Immunity
- Relaxation
- Longevity

- Fear
- Stress
- FFF
- Anxiety
- Depression
- Cortisol
- Worry
- Anger
- Addiction
- Nervousness
- Phobias
- Mood swings

THE INSULA: THE COMPASSIONATE INTEGRATOR

The insula is a region of tissue located deep inside the crevice separating the brain's temporal lobe from the frontal and parietal lobes. It has a large concentration of a special kind of neuron called von Economo neurons (VENs) that facilitate the *integration of information from many different parts of the brain.* These specialized cells are found only in a few species that form highly organized social groups: humans, apes, monkeys, whales, and elephants.

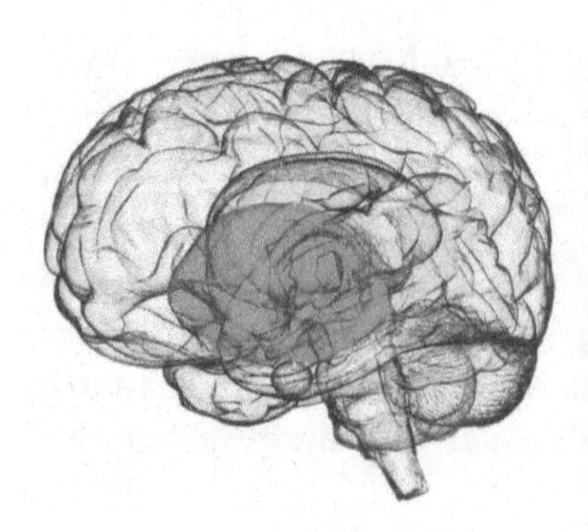

3.11. The insula: integrating information and emotions that connect.

That's because the insula is part of our caretaking circuit, one found in other mammals. Mammalian parents nurture their young, rather than laying an egg and crawling away. We also care for our family members and others of our tribe.

The insula plays a starring role in the emotions that connect us to others of the human species. These are the "social emotions," and they include pride, disgust, guilt, gratitude, resentment, embarrassment, contempt, and lust. The insula has been characterized as being "crucial to understanding what it feels like to be human."

In social settings, the VENs in the insula help us recognize emotions like disgust and guilt in other people's facial expressions. They enable us to enjoy jokes or feel slighted when a friend ignores us.

How Big Is Your Tent?

There's an old New England joke about Jess, a sullen Baptist farmer from Maine. Jess said the same prayer every night: "God, bless me and my wife. My brother Jim and his wife. Us four—no more."

To use an Old Testament metaphor, the borders of Jess's tent extended only to his immediate family. You probably have a bigger circle of people you care for: maybe the borders of your tent extend to your community, your city, or even your country.

Experienced meditators have infinite borders to their tents. Richard Davidson, and his large laboratory team at the University of Wisconsin, have tested 21 such adepts. The most experienced of these, a Tibetan monk called Mingyur Rinpoche, clocked over 62,000 hours of meditation before the age of 42.

When Davidson first hooked up Mingyur to an MRI, he gave him a number of exercises to complete, in order to determine which brain regions were engaged by each one.

When Mingyur was given the cue to engage compassion, Davidson and the other researchers in the lab's control room were stunned. The level of activity in Mingyur's empathy circuits rose by 700%. "Such an extreme increase befuddles science," wrote Davidson. On average, the adepts had 25 times the amount of gamma brain waves, the signature brain waves of a creative state of brain integration, compared to a control group. These are the waves of genius, as we'll see in Chapter 6.

The insula also gives rise to *empathy*. People who are more sensitive to emotional cues from others have greater insula activation and score higher on tests of empathy. And the insula lights up during meditation sessions, especially when the meditator is feeling kindness and compassion.

As the meditator expands his definition of connection to include other people and eventually the entire universe, he feels one with everything. In the words of a comprehensive meditation review, "the habitual reified dualities between subject and object, self and other, in-group and out-group dissipate." As he expands the borders of his tent to infinity, massive changes occur in his brain activity.

Insula Activation Benefits

- Elevated emotional states
- Motor control
- Kindness
- Compassion
- Empathy
- Longevity
- Immunity
- Happiness
- Love
- Sensory enjoyment
- Introspection
- Sense of fulfillment
- Feelings of connectedness
- Focus
- Self-awareness

- Anger
- Fear
- Anxiety
- Depression
- Addiction
- Chronic pain

As well as mediating our empathy and compassion circuits, the insula has several other functions. It collects information from a far-flung network of receptors inside our body as well as from our skin. It then stimulates feelings such as hunger that then prompt actions such as seeking food.

The dark side of this mechanism is that it can stimulate cravings for drugs, tobacco, and alcohol. Addicts show increased insula activation *even before* consuming their drug of choice. The insula also lights up when we feel pain or even *anticipate* feeling pain.

Meditators are more "in the moment" when it comes to physical pain, releasing it more quickly. They may also experience overwhelming cravings, as we'll see in Chapter 5. These are positive cravings directing them toward the ecstatic states found in Bliss Brain.

TEMPOROPARIETAL JUNCTION: THE INFORMATION SORTER

The temporoparietal junction (TPJ) is where the brain's temporal and parietal lobes connect. The TPJ also connects to the limbic system, which processes emotion and learning, as well as connecting to the visual, auditory, and somatic systems.

Our senses are being bombarded with information from the external world all the time. Our brain also has to handle all the information flooding it from the body's many organs and systems. One job of the TPJ is to sort through this avalanche of information and determine what is relevant, then package it into a coherent story for the PFC to consider and act on.

A group of investigators at the University of Illinois set out to map the parts of the brain active in emotional intelligence. They scanned the brains of 152 Vietnam veterans as they engaged in activities that required emotional intelligence, such as interacting with other people. They found that the TPJ came to life when the veterans were socially engaged.

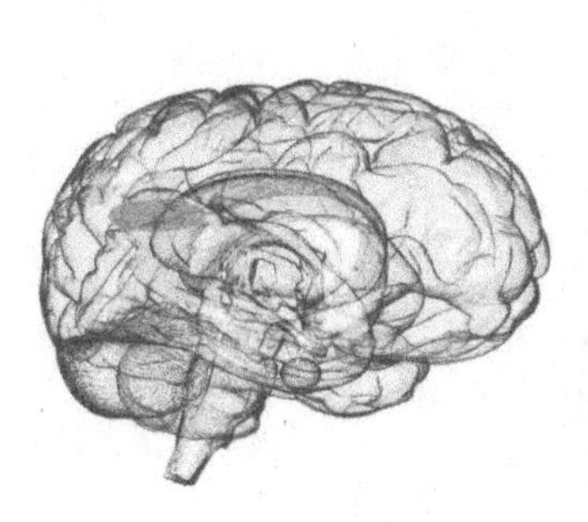

3.12. The temporoparietal junction: the information sorter.

The TPJ is active in meditation. A team of German and Spanish researchers took a group of non-meditators and gave them mindfulness training. After just 40 days, their TPJs were processing information more efficiently, while their anxiety and depression levels had plunged.

Out of Brain, Out of Body

Like Marcus Raichle's accidental discovery of the DMN described in Chapter 2, new insight into out-of-body experiences (OBEs) has emerged from Swiss neurologist Olaf Blanke's research on epileptic seizures. Searching for the source of a female patient's epilepsy, Dr. Blanke used electrodes to map her brain, pairing brain areas with the functions each controlled.

When he stimulated the angular gyrus, part of the TPJ, the patient had a spontaneous OBE. She reported to Blanke that she was looking down on herself from above. Blanke discovered that each time he stimulated that area, his patient would go into an OBE.

Blanke theorizes that in the flood of information entering the TPJ, neural pathways in epileptics might get crossed, leading to a momentary release from the borders of one's body. In meditation, this is a side effect of deliberate practice. A similar mechanism might be at work in near-death experiences (NDEs).

Physician Melvin Morse, MD, had this thoughtful comment on the relationship of these brain states to objective reality: "Simply because religious experiences are brain-based does not automatically lessen or demean their spiritual significance. Indeed, the findings of neurological substrates to religious experiences can be argued to provide evidence for their objective reality."

By activating this hub of emotional intelligence, meditation upgrades a whole host of positive qualities, including altruism, adaptability, empathy, language skills, self-awareness, conscientiousness, and emotional balance.

Temporoparietal Junction Activation Benefits

- Emotional intelligence
- Altruism
- Motivation
- Empathy
- Better relationships
- Conscientiousness
- Self-awareness
- Information processing
- Perception
- Focus
- Written language
- Spoken language
- Emotional balance

- Social anxiety
- Autism

THE PREFRONTAL CORTEX: THE COMMANDER

The prefrontal cortex (PFC) is the foremost part of the frontal lobe. Complex behaviors originate here. The PFC is also the seat of your personality, what makes you *you*. It's responsible for impulse control—moderating the emotions arising in other parts of our brains.

The PFC is the part of the brain that can say "no" when we feel like screaming at our neighbor for taking our parking spot, throttling our child when they refuse to eat their broccoli, hitting our boss in the face with a clipboard after a bad performance review, or French-kissing our best friend's hunky new husband.

It's also the part of the brain that focuses our attention and organizes our mental activity around our priorities. The PFC handles the planning of complex activities that involve many inputs such as time, space, our bodies, our resources, and other people's reactions. It can prioritize competing information.

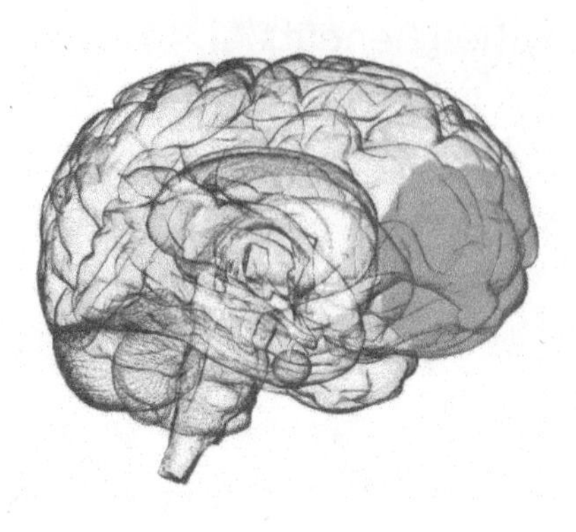

3.13. The prefrontal cortex: the commander.

In people who have the ability to focus, the PFC is able to tune out distractions. When I'm writing, for instance, I may be aware that there's a siren in the distance, that my email inbox is bulging with unread messages, that I only have a day left to pack my clothes before flying off to teach the next live workshop, and that the deadline for paying my cell phone bill is next week. But my PFC chooses to focus all my attention on the task at hand, screening out the other items clamoring for a share of my conscious awareness.

The Beavers Learn to Focus

Greg Warburton is a therapist specializing in sports performance. He grew up living and breathing sports. He completed his first marathon at the age of 27, but shortly thereafter lost a leg in a motorcycle accident.

This didn't slow him down. He became a counselor and began teaching other athletes high-performance techniques like Emotional Freedom Techniques (EFT). EFT is a super-quick method of regulating stress by tapping on acupressure points, and I often use it at the start of EcoMeditation.

I collaborated with Greg on a pioneering study at Oregon State University (OSU). In a randomized controlled trial, we measured the abilities of the men's and women's basketball teams, the OSU Beavers. After testing the athletes for the number of free throws they could successfully score, and how high they could jump, one group got EFT, while the other got a placebo treatment.

Afterward, the EFT group performed 38% better at free throws than the control group. It took only 15 minutes of EFT to produce that result.

While I was on the court observing the players, I noticed how consistent they were. Some could shoot 10 balls out of 10 into the hoop time after time.

Yet some players were consistently poor, able to make only three or four out of 10 baskets. Each time they played, they missed about as many shots as before, while the star free-throwers were just as consistent in throwing 10.

The difference certainly didn't come down to physical ability. If a basketball player can make one basket, she's certainly able to perform that task. She has the muscular strength and eye-hand coordination to dunk the ball in the basket. Theoretically, aside from fatigue, she should be able to repeat her success 10 times in a row.

The difference had to do with their *ability to regulate emotion*. I watched one low scorer carefully as he set up his shot. He was trying to concentrate, but he was distracted. The noise in the stadium, as well as the noise inside his head, was making it hard for him to focus. His PFC was not able to regulate the emotional parts of his brain.

Greg's work training the golf, baseball, and basketball teams resulted in unprecedented success for the Beavers. Soon after Greg began coaching them, the baseball team won their league twice in a row. In 2018 they won the College World Series yet again.

When they can regulate their emotions, anxiety and stress aren't able to compromise the ability of athletes. Just as long-term meditators are champion focusers, with PFCs able to suppress the emotions generated by the rattling of the DMN, athletes can be too. At the top of the game, the ability to focus can be even more important than physical strength and coordination.

The PFC is the last part of the brain to develop as we're growing up. Children don't fully develop the complex planning and decision-making abilities inherent in the PFC till they're into their teenage years, and the PFC is not fully developed till around age 25.

The PFC has several substructures. The two that are most important in meditation and focus are the dorsolateral PFC, which connects with other brain areas involved in cognition and attention, and the ventromedial PFC, which connects with the limbic system including the hippocampus and thalamus.

Ventromedial Prefrontal Cortex: The Calming Influence

During meditation, the ventromedial PFC is active. It's part of the emotional regulation circuit that we'll review in Chapter 6. Signals from the ventromedial PFC calm the emotions in the hippocampus and amygdala, giving us a break from all the feelings stirred up by the DMN. This reduces our reactivity to the outside world as well as internal emotions that distract from the meditative state.

Dorsolateral Prefrontal Cortex: The Alert Focuser

The second part of the PFC active during meditation is the dorsolateral PFC. While the ventromedial PFC suppresses emotion, the dorsolateral PFC maintains focused attention. This allows us to keep our consciousness anchored in the meditative state without our attention being distracted by input from the outside world. Once the dorsolateral PFC has made a decision, it suppresses incoming information that is now irrelevant.

By tuning out nonessentials, meditation improves our ability to pay attention. This in turn improves working memory, the ability to turn current information into long-term recollection. A study showed that students trained in mindfulness were better able to recall what they'd learned, and improved their scores on the graduate school entrance (GRE) exam.

While in this chapter we examine parts of the brain associated with specialized functions, there are few individual parts of the brain with a single purpose. The brain works together as a whole. This chapter simplifies the picture to highlight regions of the brain associated with each function, but in reality the differences in activation can be small.

Rather than on-off switches, think of dimmer switches, with a large range from light to dark. Different functions result in various brain regions getting brighter or darker, not turning on or off. And though it has all these specialized regions, the brain is orchestrated to function as a collective whole.

For instance, when we feel empathy, as meditators are trained to do, the whole limbic system including the thalamus, hippocampus, and amygdala works together with other brain regions. The "empathy circuit" also involves the insula and the anterior cingulate cortex (ACC). A large section of the amygdala is dedicated to messaging the PFC through the ACC. The greater the degree of empathy experienced, the greater the activation of the empathy circuit.

The ventromedial PFC and the dorsolateral PFC are two relatively small and specialized parts of the PFC. In meditators, the "selfing" parts of the PFC go offline during practice. Brain scans of meditating monks show that the parts of the PFC that construct our personalities go dark, with energy usage dropping by as much as 40%—the "transient hypofrontality" noted by neuroscientists in Chapter 2. Newberg finds that many different types of practitioners "get out of their heads," from Brazilian shamans to Pentecostals who are "speaking in tongues."

While we're in meditation, we lose our identification with our stories about ourselves and the world. For a while, we stop selfing. We forget I-me-mine. The bonds that keep our consciousness stuck in ego, in looking good, in remembering who we like and dislike, in playing our roles—and all the suffering that accompanies these things—are loosened. That frees us up to enter nonlocal mind, and *bond with a consciousness greater than our local selves*. Newberg describes it this way: "The person literally feels as if her own self is dissolving. There is no 'I'—just the totality of a singular awareness or experience."

The paradox of enlightenment is that we have to *lose our personalities to find bliss*. While the thinking abilities of the PFC are our biggest asset in everyday life, they're our biggest obstacle to experiencing oneness. It's the ego that separates us from the universe, and when it goes offline, we join the mystery.

Prefrontal Cortex Regulation Benefits

- Emotional regulation
- Executive brain function
- Goal-setting
- Complex reasoning
- Appropriate behavior
- Intellectual intelligence
- Orchestrating ideas
- Happiness
- Decision-making
- Willpower
- Impulse control
- Focus
- Well-being
- Meaning and purpose
- Resilience
- Fluid intelligence
- Attention
- Cognitive flexibility
- Working memory
- Self-control
- Prosocial behavior
- Long-term thinking
- Complex planning

- Anxiety
- Addiction
- Distraction
- Depression
- Selfing
- Reactivity to stress
- Pain
- Mind Wandering
- Startle response
- ADHD
- Alzheimer's
- Cognitive decline
- Cortisol
- Inflammation
- PTSD
- Aging
- Chronic fatigue

THE STRIATUM: THE EMOTIONAL THROTTLE

The striatum is part of the basal ganglia. These ganglia are called "basal" because they're near the bottom of the brain and connect to the brain stem and spinal cord through which signals travel throughout the body. *Striatum* is Latin for "striped." It's composed of white matter with stripes of gray matter running through it.

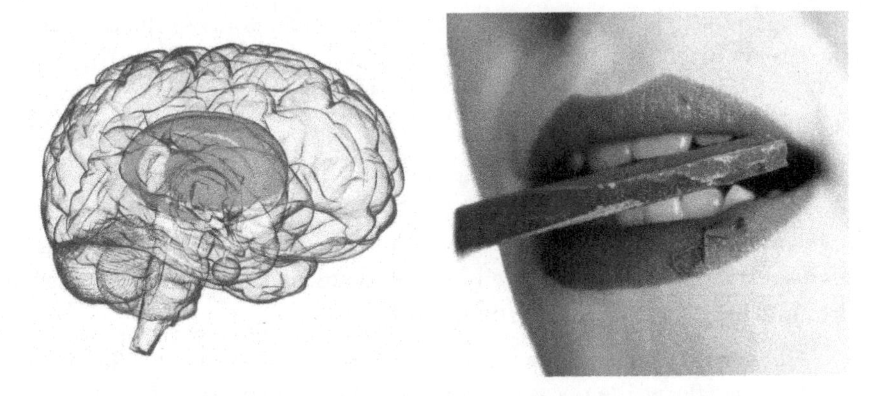

3.14. The striatum: the regulator of emotions and cravings.

It contains several substructures important to emotional regulation. One of these is the caudate nucleus, a C-shaped curve of tissue near the base of the neocortex. The caudate nucleus plays an important role in the processing and storage of memories, especially as they relate to new learning and language. Like the hippocampus, it draws on past experiences to *interpret the importance of incoming information.*

It moderates stress by serving as a gateway of information between the cortex and the limbic system, and it can *throttle the transmission of stressful information* between the two. If it's engaged, it regulates the flow of worried emotions. If not, it can allow them to flow too freely, leading to unchecked anxiety.

Once your PFC has formulated a goal, it communicates it to the basal ganglia, which in turn formulate the appropriate actions for achieving the goal. The basal ganglia also inhibit movement that does not serve the goal. The result is fluid, coordinated movement. In Parkinson's disease, the basal ganglia deteriorate. This results in tremors, muscular rigidity, and a loss of coordination.

The caudate nucleus is active when we're paying attention and learning. It helps us remember what we've learned, so we can do the same thing on autopilot next time. When, for instance, you can ride a bike without having to think actively about every part of the task, the other parts of your brain are freed up. That enables you to enjoy the scenery and stay safe in traffic.

Peak performers engage the caudate nucleus to *keep themselves in states of flow.* They "throttle down" emotions such as fear and anger so they can engage fully with the task at hand. When meditators use the caudate nucleus to throttle back their negative emotions repeatedly, *the process becomes automatic*, like riding a bike. This makes them resilient in the face of stress.

We encountered the nucleus accumbens, another important part of the striatum, in Chapter 2. It's associated with rewarding experiences and the reinforcement that reward produces in the brain. It's activated by pleasurable experiences, during which it secretes large amounts of dopamine, the brain's primary reward neurotransmitter.

This reward system plays a role in addiction. Drugs like alcohol, heroin, and cocaine trigger the release of dopamine in the nucleus accumbens. It also kicks in when you find a $20 bill on the beach, have an orgasm, or help yourself to a generous portion of cherry pie.

But when a meditator contemplates altruism, her nucleus accumbens lights up. She gets the same rush of dopamine that an addict gets when he sniffs a line of cocaine. Same for the chocoholic unwrapping her Ferrero Rocher truffle. Meditation makes meditators feel good *using the exact same neurotransmitters and brain regions active in the addict*, as we'll see in Chapter 5. This reward system explains why long-term meditators maintain a regular practice. They're addicted to feeling wonderful!

Striatum Regulation Benefits

- Adaptive behavior
- Motivation
- Appropriate goals
- Emotional regulation
- Appropriate emotionality
- Long-term memory
- Motor coordination

- Anxiety
- Autism
- Obsessive-compulsive behavior
- ADHD
- Anxiety
- Addiction
- Distraction
- Depression

THE PONS: REGULATOR OF THE BEAST

The pons is part of the brain stem, the oldest part of the brain. It's about the size of your fingernail, and it plays a role in many physiological functions that happen continuously outside of your conscious control. The pons regulates your breathing, including how many breaths per minute you take and how deeply you breathe. It automatically adjusts these when needed, such as when you engage in physical exertion or lie down on the couch for a nap. It also plays a role in taste, hearing, balance, and deep sleep.

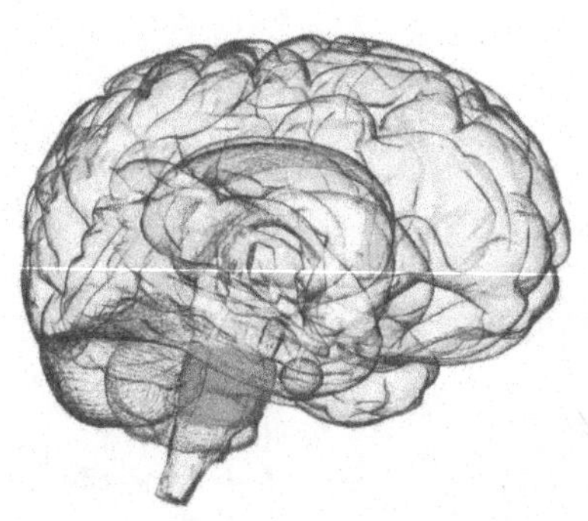

3.15. The pons: the regulator of the beast.

The pons is active during meditation, as we breathe deeply and regularly. It's associated with the production of delta and theta waves in the brain, which research shows turns on a host of healthy processes in your cells. These include increased stem-cell production and the repair of skin, bone, muscle, nerves, and cartilage. These brain waves also lengthen our telomeres, the most reliable marker of longevity.

A remarkable ability of humans is that *we are able to activate or deactivate all of these brain regions by consciousness alone.* We can shift our thoughts deliberately with meditative practices or simply by focusing on different stimuli. The brain responds accordingly. We'll see the extraordinary neural effects of this superpower of "selective attention" in Chapter 6, and the evolutionary implications in Chapter 8.

Pons Activation Benefits

- Quality REM sleep
- Cell repair
- Longevity
- Energy
- Cell metabolism
- Melatonin
- Delta brain waves
- Theta brain waves
- Dream frequency and quality
- Lucid dreaming

- Insomnia

To the Brain, Imagination Is Reality

For thousands of years, sages have assured us that our minds create our reality. In Proverbs 23:7, the poet tells us that, "As a man thinketh in his heart, so is he." Two thousand years ago the Buddha said, "We are what we think. All that we are arises with our thoughts. With our thoughts, we make the world." Now neuroscience is showing us how true this is.

An ingenious study measured how our brains respond to scenarios that exist *only in our imaginations*.

A research team at the University of Colorado at Boulder took 68 people and gave them a mild electric shock accompanied by a sound. They were then divided into three groups.

The first group heard the sound repeatedly, though this time without the shock. The second group imagined the sound in their heads repeatedly. The third group imagined the pleasant natural music of rain and birds.

The group imagining the sound showed the same brain activity as the one actually hearing the sound. Two brain regions, the ventromedial prefrontal cortex and the nucleus accumbens, lit up. As we've seen, the first regulates emotions like fear in the limbic system, while the second processes reward and aversion.

Later, people in the "rain and birds" group were still afraid of the sound even when it was repeated many times without the shock. But those in the group that heard the real sound, *as well as those imagining it*, unlearned their fear. In neuroscience, this revision of reality is called "extinction learning." In Chapter 7, we'll see how the ability to extinguish fear produces resilience and posttraumatic growth.

The study's lead author, Marianne Reddan, said, "Statistically, real and imagined exposure to the threat were not different at the whole brain level, and imagination worked just as well."

Tor Wager, director of the Cognitive and Affective Neuroscience Laboratory at the university and co-author, said, "This research confirms that imagination is a neurological reality that can impact our brains and bodies."

The take-home message is that *to our brains, our thoughts are reality.* When our DMN leads us to catastrophize and ruminate about the bad things in life, we activate the same brain regions that are active when the bad stuff is actually happening. That in turn will "interrupt the brain's ability to perform well on every level," according to Wager.

But when we meditate or use "selective attention" to focus on pleasant realities filled with kindness, joy, compassion, and creativity, we use our consciousness to light up the corresponding brain regions. This turns on hundreds of genes that regulate our immunity, energy levels, insulin, stress, antiaging molecules, and the health of our cells' mitochondrial powerhouses.

GETTING ADDICTED

Flip through the Benefits boxes in the preceding pages and you'll see the astonishingly wide range of gains you make by turning on your Enlightenment Circuit. Meditation doesn't just regulate one or two brain regions. It has massive effects throughout the brain and, in turn, on the body. You feel very different physically. Your emotions shift from negative to positive, propelling you to euphoria.

The DMN's activity means that the brain's energy usage is very stable throughout the day, varying by less than 5%. So the 40% drop in PFC and parietal lobe activity that occurs during meditation represents an enormous shift. The meditator feels completely different when this threshold is reached. She enters a sublime new dimension of reality—not just in her subjective experience, but also in her objective brain function.

The shutdown of these two brain regions liberates the meditator from selfing, from the nagging voice of dissatisfaction, and from attachment to the here and now. Done repeatedly, the brain reorganizes itself around this new state. Richard Davidson describes this as "a transformation that dramatically ups the limits on psychological science's idea of human possibility."

The transcendent experiences people have once they learn to induce Bliss Brain are so profound that they are addicting. Their brains are flooded with the same reward neurotransmitters as those of a hard-core addict, as we'll see in Chapter 5. This makes meditators want to repeat the experience and make meditation a regular practice, rather than an occasional dabble.

The beneficial brain states that meditation produces are reason enough to practice it. Coupled with the improvements it produces in emotion, the healing effects it has on the body, and the transcendent nonlocal connections it provides, meditation is able to produce a pervasive sense of well-being. It is the highest-leverage point you have over your quality of life—spiritual, mental, emotional, and physical. Taken seriously and practiced regularly, meditation unlocks the door to a wonderfully happy life. Once you taste Bliss Brain, there's no going back.

DEEPENING PRACTICES

Here are practices you can do this week to integrate the information in this chapter into your life:

- **Activating the Enlightenment Circuit at the Start of Each Day:** Each morning, listen to the free 15-minute EcoMeditation on activating the Enlightenment Circuit you'll find in the Extended Play Resources section below. Do this first thing in the morning, before the thoughts of the day rush in to occupy your mind.
- **Replacing Annoyance with Empathy:** Notice when you're feeling annoyed by someone. Visualize what form the annoyance takes in your body. Where in your body is it located? What color is it? What shape is it? Imagine pulling it out of your body and setting it down nearby. Then, tune in to the way you felt during that morning's meditation. Imagine pulling that empathetic energy into your body, and putting it in the same place. Direct the empathy toward that nearby annoyance. It usually dissolves.

- **Imagination Break:** Give yourself a 5-minute "imagination break" before noon each day. Use it to imagine the most positive outcome in a single area of your life.
- **Whole-Brain Sensory Reinforcement:** In your personal journal, write down what you've imagined in great detail, and engage each of the five senses. What does your imaginary positive outcome smell, taste, touch, sound, and look like?

EXTENDED PLAY RESOURCES

The Extended Play version of this chapter includes:

- Dawson Guided EcoMeditation: Activating the Enlightenment Circuit
- Video of Brain Regions Active in Meditation
- Audio Interview with Andrew Newberg, PhD, Author of *How Enlightenment Changes Your Brain*
- Dawson Hay House Video on Activating the Brain's Enlightenment Circuit

Get the extended play resources at: BlissBrainBook.com/3.

CHAPTER 4

THE ONE PERCENT

The top 1% of the world's wealthy control more than 50% of all wealth, according to Credit Suisse's global wealth report. In the United States, the 1% own more wealth than the bottom 90%. The number of millionaires in the world has tripled in the decades since 2000. And the amount of the world's wealth controlled by the bottom 50% of the global population? Under 3%.

These inequalities are more than numbers. They are fuel for high emotions and mass social change. They have led to the rise of populist political movements and propelled a variety of unlikely candidates into power. The difference between the top 1% and all the rest gets our attention.

So much for money. Let's now consider something infinitely more valuable: happiness. Specifically, the happiness found in Bliss Brain.

Here we also find huge inequalities. Historically, Bliss Brainers are a tiny percentage of the population. Few even attempt the journey to enlightenment, and of those who seek Nirvana, even fewer attain it. When a rare spiritual genius, such as Jesus or Buddha, reached that pinnacle, the event was so significant that it changed the entire course of world history.

WITHDRAWING FROM EVERYDAY LIFE

The lives of the great spiritual masters of history inspired others to follow their example. But like the saints, these aspirants could not reach enlightenment in the everyday world, with its demons and distractions.

So for thousands of years, those committed to the spiritual path went to special places such as hermitages, wilderness retreats, monasteries, and convents. They exiled themselves from ordinary society in order to pursue nonordinary states of consciousness. They couldn't achieve Bliss Brain amid the hubbub of society, so they turned their backs on it.

The rest of society stayed in ordinary consciousness, driven by the desires and demons of the Default Mode Network (DMN). In my book *Mind to Matter,* I call this survival orientation "Caveman Brain." It's hard to find Bliss Brain when surrounded by Caveman Brain, and pulling yourself out of that environment and into a sacred space is usually a prerequisite for enlightenment.

What percentage of the population undertook the journey? No census of enlightenment seekers is possible, but one proxy is the number entering religious seclusion. In the early 1300s, England had a monastic population of about 22,000, with another 10,000 in other religious occupations.

4.1. Traditionally, withdrawal from everyday life was required to attain enlightenment. See full-color version in center of book.

THE ENLIGHTENMENT-SEEKING MINORITY

That was a total of 1% of the population of that time. Estimates are similar for other medieval European countries. Those who actually reached enlightenment were just a fraction of these aspirants.

In Tibet, traditionally one of the most religious countries, surveys from the past 500 years estimate that around 12% of the population has been monastic. But today in Thailand, a more typical Buddhist country, monks and nuns comprise only 1%.

History's best estimates thus tell us that the spiritual path has been walked by about 1% of the entire population. These are the people who have made enlightenment their wealth.

These "One Percenters" focused on studying and experiencing Bliss Brain. As they laboriously acquired knowledge of how to do this, they recorded and curated their discoveries in their cloisters and monasteries. Meditation and the steps of initiation were carefully guarded secrets reserved for initiates.

Their paths, like those of Jesus and Buddha, were not easy. The writings of the initiates describe a long and arduous spiritual journey. The journey from ordinary consciousness to Bliss Brain had many stages, with difficult and sometimes painful initiations at each step.

Over 2,000 years ago, the prince Gautama, later to become the Buddha, pursued enlightenment. He faced many obstacles. He studied with two learned Brahmins and followed all the rituals they recommended. He was the "holiest of the holy," taking initiatory practices to extremes of self-deprivation. Eventually, he became so emaciated that a passing girl named Sujata took him for a ghost. She restored his strength with milk and rice pudding.

Yet after 6 years of effort, he still found himself far from enlightenment. Not only had his single-minded determination failed, his effort had in itself become an obstacle.

He sat under a fig tree for 49 days, wrestling with doubt and despair. That's when he broke through to the "middle way" between self-denial and self-indulgence, and found enlightenment. He experienced the source of true wealth, the most valuable secret. His followers then encoded his teachings of the Four Noble Truths and the Eightfold Path into the core tenets of Buddhism. These show the way of the One Percenters to the other 99% of us.

4.2. The eightfold path of the Dharma Wheel in sculpture.

Steps of Initiation

In ancient Greece and Rome, the Eleusinian mysteries were regarded as the most important of initiatory ceremonies. They were believed to contain such power that they could unite humans with the gods, while conferring superhuman gifts and the promise of the survival of the initiate's consciousness beyond death.

Roman historian Marcus Tullius Cicero wrote: "For among the many excellent and indeed divine institutions which your Athens has brought forth and contributed to human life, none, in my opinion, is better than those mysteries. For by their means we have been brought out of our barbarous and savage mode of life and educated and refined to a state of civilization; and as the rites are called 'initiations,' so in very truth we have learned from them the beginnings of life, and have gained the power not only to live happily, but also to die with a better hope."

There were two separate ceremonies. The "Greater Mystery" was observed every 5 years and the "Lesser Mystery" every year. Steps of initiation included bringing sacred objects to the Acropolis, washing in the sea, harvesting ceremonial plants, proceeding along a sacred

4.3. The Eleusinian mysteries. See full-color version in center of book.

route, feast days, fast days, drinking an LSD-like psychedelic, lighting a sacred fire, sacrificing a bull, and pouring prescribed solutions from sacred goblets.

The penalty for speaking publicly about the most holy stages of the rite was death.

Whether it's the Catholic ritual of the Twelve Days of Christmas, the Mysteries of Isis of ancient Egypt, the fasting ceremonies of the Muslim month of Ramadan, or the Orisha rites of the African Yoruba tribe, humans have for millennia formalized the stages of initiation into altered states of consciousness.

Mystical states were regarded as the most treasured of experiences. People were willing to give up almost anything to attain them. They recognized them as so special that those who reached the mountaintop were revered forever. The practice of meditation was taught as the entry point to the mystical experience. It marked the path to the enlightenment enjoyed by the One Percenters.

That's the promise drawing so many people into meditation today. We know there's an extraordinary state at the end of the journey. Perhaps a flash of deep insight, enlightenment with a small *e*. Or even an experience that drops us into bliss forever, Enlightenment with a big *E*. We'll chart the increasing number of meditators in Chapter 8.

CONFESSIONS OF A FAILED MEDITATOR

That's why, as a teenager in the 1960s, I dabbled in meditation. I was desperate to escape the anxiety and depression that had haunted the first decade of my life.

One day, on the way to meet friends in a hotel lobby, I walked past a full-length mirror. I stopped and I stared at the tall gangly youth in front of me. Shoulder-length hair, parted in the middle. Sky-blue bell-bottom trousers. A tie-dyed man-purse slung over my shoulder.

As I examined my features, the words popped into my head: "That's the saddest face I've ever seen."

That was me, and most people I knew. But amid the sea of miserable people around me, there were one or two who seemed genuinely happy. I thought, "Maybe they're just masquerading, their smiles hiding secret desperation."

Yet perhaps they were genuinely serene, untroubled by the anxiety, depression, self-hatred, and fear that was my normal internal state. "Could meditation be my escape route?" I wondered.

By 16 I was living in a spiritual community. We cultivated organic vegetables, ate vegan, and studied what sage Aldous Huxley termed "the perennial philosophy" that underlies all religions. I discovered the reflections of Dr. Paul Brunton, a seeker who between the 1920s and 1950s traveled all over the world seeking out its spiritual traditions and writing perceptively about enlightenment in a famous set of notebooks. I devoured the adventures of Lobsang Rampa, a Tibetan Buddhist monk who wrote a series of bestsellers in the 1970s.

I kept a journal in an effort to understand my own suffering. I could not bridge the gap between the elevated states described by these mystics and my own my endless parade of problems and challenges. I understood the bliss the mystics described as well as I understood Neil Armstrong's experience of walking on the moon. It happened, but not to me.

In meditation class the guru assured us, "Meditation is easy. All you have to do is still your mind."

"Still my mind!" I screamed inside. "Are you nuts? That's impossible!"

Closing my eyes made my hyperactive thoughts race even faster. As the TPN went dark, my DMN went into overdrive. Over the next few years, whenever I tried meditation, I came face-to-face with the demon. The serenity of the saints eluded me. This gave me little incentive to stick with a regular practice.

The spiritual community itself was a hot mess of hypocrisy, misguided zeal, back-stabbing, manipulation, ignorance, and inept leadership. This gave my DMN plenty to chew on. Eventually the group imploded in a stew of sexual and financial scandals.

I went forth into the world disillusioned, utterly unequipped to understand money, relationships, health, and career. Struggling with those occupied my attention for the next couple of decades, and closing my eyes in meditation, to wrestle again with the demon, took a back seat. The serenity of the saints was a pipe dream.

4.4. Dún na Séad

SEEKING CLEAR VISION

When I turned 30, I felt an inner calling to go on a long retreat. I wound up all my business projects before Christmas and went away to

Ireland for a month. I booked a cottage in a remote hamlet called Dún na Séad, in County Cork. That's as far away from Dublin as you can get without splashing into the Atlantic.

I sought out ancient stone circles, placing my hands on the cool, mossy rock and visualizing the civilizations that had set up these sacred sites. I spent hours walking the Irish cliffs, with the Atlantic beating wildly below. I hiked through the forest trails of the Emerald Isle. I filled up a whole journal with my reflections.

After a couple of weeks, I felt my mind calming down. I meditated occasionally, and sometimes felt one with the universe. As I connected with nonlocal mind, I felt love filling me despite the chaos of my thoughts. Whenever I was "tuned in," I felt a pressure in the center of my forehead. Like a clearing of light in a dark forest, I experienced patches of inner peace despite the turmoil of my life. After that I meditated occasionally though not regularly.

THE DAY EVERYTHING CHANGED

Fast-forward a decade and a half. I was 45 and sitting at a Mexican restaurant called El Capitan, surrounded by murals of giant sombreros and three-armed cacti. Opposite me sat Salli, my life coach.

"I'm overwhelmed," I told Salli. "I'm a single dad with two young children. School begins at 8 A.M. and we have to leave the house at 7:30 A.M. That means dragging my sleepyheads out of bed at 6:45 A.M. and myself at 6:30 A.M.

"I have two businesses, one going downhill and one thriving. I have no time for myself. For the gym. For leisure. For play.

"Every week I write in my journal. Despair. Hopelessness. Emotional agony. Physical pain. Nightmare girlfriends. No self-confidence. Money struggles. Shattering disappointment.

"I've sat at the feet of masters. I've read all the inspirational books and taken the usual workshops. Nothing makes much of a difference.

"When I pick up an old journal from 5 or 10 years back, I read the same old, same old. Only the dates are different. I'm on a hamster wheel, running hard, and happiness just gets further away."

Can you relate?

Salli asked me whether I meditated regularly.

I responded bitterly, "I have no time for meditation!"

Like an icy cloud of breath on a cold winter morning, the words hung in the air in front of me. I stared at them.

With utter certainty, I knew that *daily* meditation was the next step in my personal growth. At that moment, I resolved to meditate daily. *No matter what.* The next day I set my alarm clock for 5:30 A.M.

That's a painful number for me. I'm a natural night owl, preferring to rise at 10 A.M.

4.5. Early HeartMath HRV monitor. Now the technology is on smartphones, but this early miniaturized device trained thousands of people to achieve heart coherence.

But at 5:30 A.M. the next morning I meditated. And the next day. And the next.

Day by day, I persisted. I actually began to look forward to getting up early.

At the end of that year, as I looked over the entries in my journal, I realized that my life was changing. Like green shoots poking their way through the barren gray wasteland, positive trends were appearing. Happy kids. Money saved. A loving relationship. Eating healthy. Regular exercise. Positive friends. Setting goals and meeting them.

That moment at El Capitan was a pivot point in my life. Happiness was on the horizon.

A half-finished book that a publisher had been wanting me to complete for 10 years was suddenly done. I completed my doctorate, begun at Baylor University 20 years earlier. I got seriously into science.

I bought an EmWave monitor and practiced entering a state of heart coherence. I learned EFT tapping. I investigated neurofeedback. I realized that technology could help us attain and understand the mystical state.

REVERSE ENGINEERING ENLIGHTENMENT

Mystics have described their experiences in an attempt to map the path for the other 99% of us. They use ecstatic poetry or luminous prose.

These soaring texts are in a realm completely separate from the reductionistic language of science. Enlightenment has been viewed as a mystical phenomenon belonging to the most esoteric levels of spirituality, unconnected to the domain of science.

Then along came technology. The EEG was invented by German physician Hans Berger in the 1920s, after a psychic experience that I describe in my book *Mind to Matter*. By the 1960s, the idea occurred to pioneering researchers that they could use EEGs to peek inside the brains of meditators.

A British engineer named Maxwell Cade hooked up enlightened people to a simple EEG he called the Mind Mirror. Other pioneers such as Robert Becker studied the brain-wave patterns of adepts from many different traditions: Buddhist monks, Pentecostal faith healers, Taoist masters, Sufi dervishes, qigong practitioners, Hindu sadhus, and Jewish kabbalists.

After recording thousands of EEGs, researchers developed a profile of the subjects' brain function. Every time an adept went into an altered state, their brain waves changed. Their ratios of beta, alpha, theta, and delta altered, forming a unique and distinctive pattern.

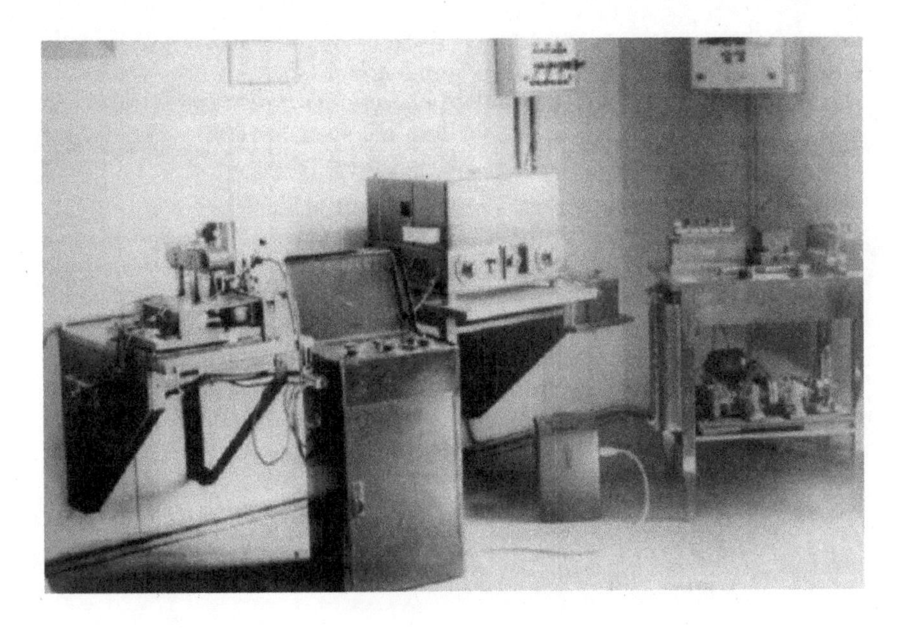

4.6. Hans Berger's EEG.

Cade called this profile the Awakened Mind. This enlightenment pattern was completely different from the neural signaling seen in ordinary states of consciousness. Scientists could determine whether or not an adept was in Bliss Brain simply by looking at their brain-wave ratios.

Not only that, but the Awakened Mind pattern looked the same regardless of spiritual practice or religious faith. A Catholic nun might be praying to Mother Mary, a Zen practitioner meditating on the Empty Circle, a Sufi chanting the names of Allah, a qigong master "gathering the qi," but their EEG readouts all looked similar. They had the characteristic waves of the Awakened Mind: small beta, big alpha, and high theta and delta.

The astonishing truth emerged: Enlightenment is a formula.

And the formula wasn't the exclusive property of spiritual adepts. Cade's student and successor, Anna Wise, began to hook up extraordinary creators and high achievers in "flow" states. Artists, CEOs, musicians, athletes, scientists. She found that many had the Awakened Mind pattern too, with lots of alpha waves and big theta and gamma. Cade and Wise, realizing that the formula was repeatable and teachable, began to train ordinary people to acquire the Awakened Mind pattern.

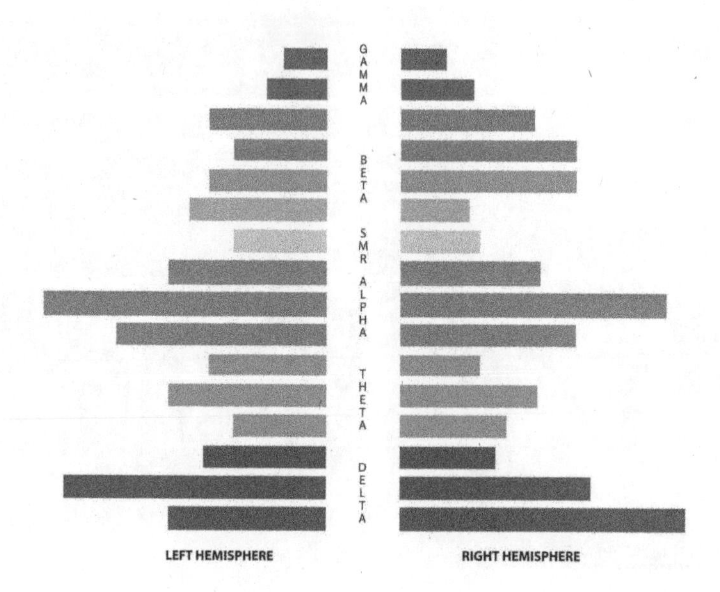

4.7. The Awakened Mind pattern. See full-color version in center of book.

As new technologies like HRV monitors, SPECT scans, PET scans, and fMRIs came along, the ability of scientists to measure these states became steadily more refined. Scientist-mystics "reverse engineered" enlightenment. By simply looking at a brain scan, anyone trained to recognize these brain-wave ratios can "read" the person's spiritual state.

This democratized enlightenment. Armed with this knowledge, novices could be trained to acquire Bliss Brain. It went from the exclusive domain of the One Percenters to a commodity that anyone could attain, given sufficient dedication, focus, and practice.

CAVEMAN BRAIN VERSUS BLISS BRAIN

I often ask two questions at the start of my live workshops. One: "Who has taken a meditation workshop, read a book, or enrolled for an online class?" Virtually every hand in the audience goes up.

Two: "Who has a consistent daily practice?"

A few hands go up—while everyone else develops a sudden interest in the wood-grain patterns in the floor.

People know they *should* meditate. They *want to* meditate. They *try to* meditate. Yet they fail to meditate. Why?

The neocortex of humans, that big bulge above your ears, is a relatively recent evolutionary innovation. The first neocortical tissue appeared in mammals 200 million years ago, but only grew huge quickly when humans learned to use tools 3 million years ago. The human neocortex went on to learn to distinguish past from future, create abstractions like music, art, and poetry, and invent language and civilization. It now accounts for 76% of the gray matter of the brain.

Your cerebellum and brain stem, seat of your survival instincts, are 400 million years old. They've been doing their job of keeping you and your species safe for 20 times longer than the humanoid neocortex has been around.

When these ancient parts of your brain are active or rehearsing the next disaster using the DMN, they effortlessly hijack your attention. You try to meditate and repetitive negative thinking takes over. In the cage match between Caveman Brain and Bliss Brain, Caveman Brain always wins. Survival is a more important need than happiness or self-actualization. You can't self-actualize if you're dead.

In 2015 the US National Institutes of Health estimated that less than 10% of the US population meditates. One of the primary reasons for this is that meditation is hard. Most people who start a meditation program drop out.

GETTING THE BEST OF ALL WORLDS

When writing my first best-selling book, *The Genie in Your Genes,* I experimented with many schools of stress reduction and meditation. Heart coherence. Mindfulness. EFT tapping. Neurofeedback. Hypnosis.

One day I had a Big Idea: *What happens when you combine them all?* I began playing with a routine that did just that. Here's what I came up with:

First, you tap on acupressure points to relieve stress. **Second**, you close your eyes and relax your tongue on the floor of your mouth. This sends a signal to your vagus nerve, which wanders all over your body, connecting all the major organ systems. It's the key signaling component of the parasympathetic nervous system, which governs relaxation.

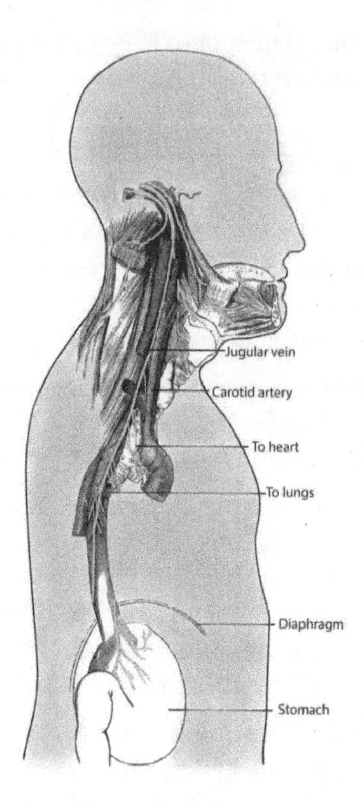

4.8. The vagus nerve connects with all the major organ systems of your body. See full-color version in center of book.

Third, you imagine the volume of space inside your body, particularly between your eyes. This automatically generates big alpha in your brain, moving you toward the Awakened Mind. **Fourth**, you slow your breathing down to 6 seconds per inbreath and 6 seconds per outbreath. This puts you into heart coherence.

Fifth, you imagine your breath coming in and going out from your heart area, and you picture a sphere of energy in your heart. **Sixth**, you send a beam of heart energy to a person or place that makes you feel wonderful. This puts you into deep coherence. After enjoying the connection for a while, you send compassion to everyone and everything in the universe.

Feeling universal compassion produces the major brain changes seen in fMRI scans of longtime meditators. As we'll see in Chapters 6 and 8, compassion moves the needle like nothing else.

At this point, most people drop into Bliss Brain automatically. They're in a combination of alpha, heart coherence, and parasympathetic dominance. They haven't been asked to still their minds, sit cross-legged, follow a guru, or believe in a deity. They've just followed a sequence of simple physical steps.

After a few minutes of universal compassion, you again focus your beam on a single person or place. You then gently disengage and draw the energy beam back into your own heart. **Seventh**, you direct your beam of compassion to a part of your body that is suffering or in pain. You end the meditation by returning your attention to the here and now.

I first tried this combination with an audience of about 200 people in Toronto, Canada, in 2009. To my amazement, they all came into coherence after a few minutes, their breathing synchronizing spontaneously. Afterward many told me it was the first time they'd ever been successful at meditating.

I realized that while each of the seven practices are powerful in themselves, *in combination they reinforce each other.* Since they are physical and mental stimuli that *require no prior training or belief,* anyone can reach an elevated state *the first time* they try. I named the method "EcoMeditation." If you'd like to read the studies validating each component of the practice, you'll find them in the References section at the end of this book.

THE FAST TRACK TO ELEVATED STATES

At the end of the sixth step, I have people disengage their beam of heart energy, and bring it back into their own hearts. This is important because we need to understand the boundaries between us and other. It's delightful to blend energy with other people, but it's vital to be able to disengage and re-inhabit your own energy space.

When people open their eyes, I have them look around and notice the objects in their environment. I might ask, "Notice the smallest green object you can see" or "What's the biggest round object?" That's because we lose our sense of self in Bliss Brain, and it's important to come back fully to local reality. Life goes on. Chop wood, carry water.

The EEGs of people using EcoMeditation for the first time in a weekend workshop are fascinating. In Mind Mirror analyses performed by Awakened Mind trainer Judith Pennington, she observed, "In two days, many participants acquired elevated brain states normally found only after years of meditation practice." They've been initiated into the mystery of the One Percenters *without practice, belief, preparation, religion, initiation, or 10,000 hours.*

The experience is so powerful and so easy that it motivates many to continue, and in Chapter 5 we'll see how compelling it can be. At the end of a workshop, when I ask, "Who will commit to a daily meditation practice?" about 90% of hands shoot into the air. Bliss Brain is addictive, and once you've experienced it, you want more.

The 30-Day Experiment

by Marie-Beth Stuckley

I've taken a bunch of meditation classes over the years. But I could never make it stick. I felt ashamed of this and awkward whenever I heard people talk about meditation.

Then a friend of mine sent me a link to Dawson's EcoMeditation tracks on Insight Timer. I listened to one and I was in a deep state right away.

So I made a decision to do it every day for 30 days—after all, the tracks are only 15 minutes long. I began to go deeper and deeper. I thought it would be hard and that I would be tempted to skip days. But I enjoyed it so much I didn't need persuading; it became part of my morning routine.

I started to see changes. I'm less impatient with my husband and kids. I'm a therapist, and after all these years I felt close to burnout. That's gone away and I'm much more positive about my job and my life.

I'm now on day 46 out of 30—LOL! And I don't see any reason to quit. I've now shared that link with my clients and friends, because if something can help you so much with so little effort, it's worth doing. It's helped many of my clients feel calmer too.

I'm now taking live classes and training in these methods, because I want my clients to get the benefit of this new science. I think it's going to change the whole profession of psychology in the next few years.

READING YOUR OWN BRAIN WAVES

The early EEGs used equipment the size of a pickup truck. They cost hundreds of thousands of dollars to construct. It took teams of specialists to hook up subjects, produce recordings, and interpret the results. EEGs were found only in the most advanced university neuroimaging labs.

By the 1980s, the machinery had shrunk to the size of a backpack, and in the 1990s, the tech was ported to laptops. Nowadays, you can read your brain waves or HRV off your cell phone. The EEG data collection device sending the signal to your phone weighs a few ounces and costs a few hundred dollars. In a few years, they'll be integrated into the fabric of your shirt. In Chapter 8, we'll see how this is going to shape the future of the world.

I've now hooked up many people using a simple EEG recording headband called the Muse. Though it lacks the sophistication of lab devices, it's capable of producing a signal that can be interpreted by someone without any scientific training. Rather than the 19 electrodes commonly used in labs, it uses just 4 electrodes. But that's enough to measure activity in the prefrontal cortex (PFC) and the temporoparietal junction (TPJ). The Muse is so comfortable that after a few minutes, wearers typically forget it's there.

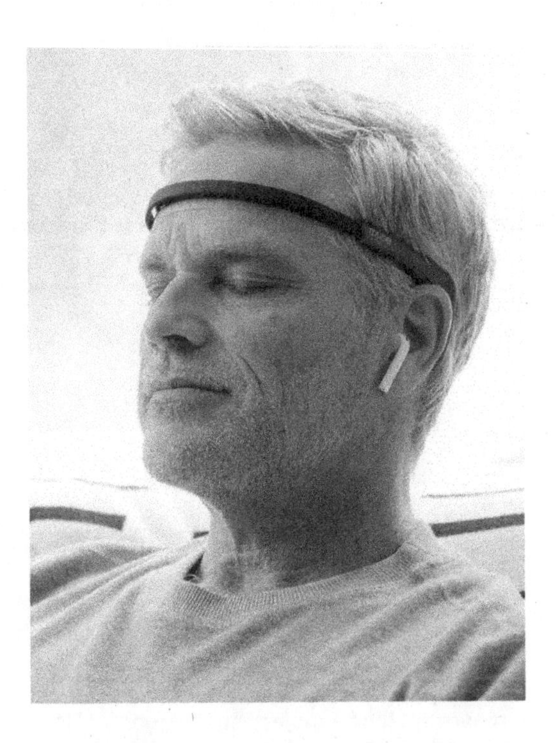

4.9. Person wearing a Muse headband.

The app that comes with the Muse trains you in how to meditate using the feedback you receive from the device. You can now be your own EEG neurofeedback technician, no lab or PhD required. Is that cool or what!

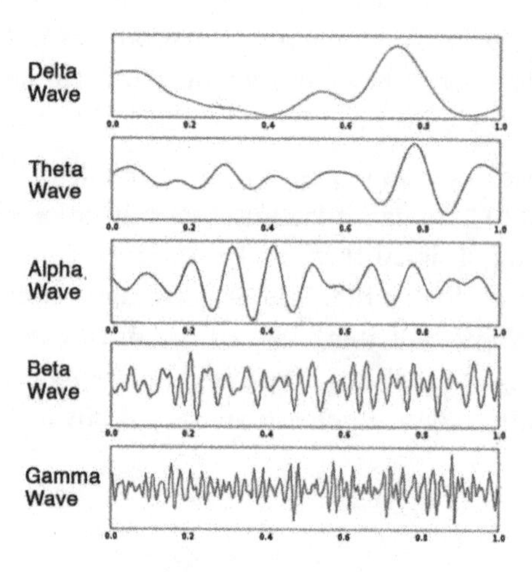

4.10. Brain-wave frequencies.

Frequency is one of the two ways we measure brain waves. Frequency is the type of brain wave, such as alpha, beta, or delta. The second way, amplitude, is how strong those brain waves are.

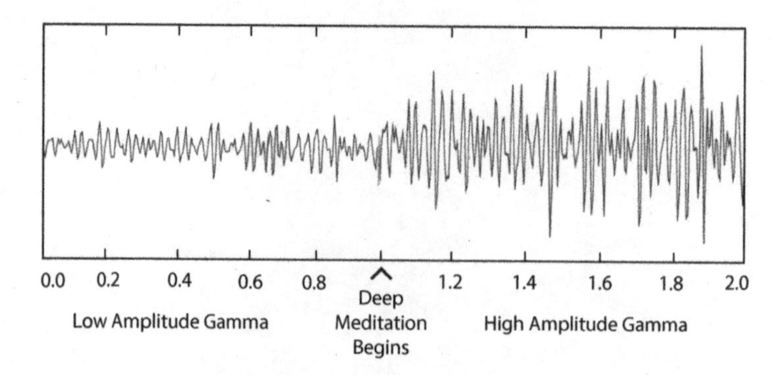

4.11. Brain-wave amplitudes.

When a researcher says that a subject has "bigger" or "increased" or "greater" or "more" of a brain wave, we mean greater amplitude. When we say that a brain wave "shrinks" or "reduces" or "lessens," we mean reduced amplitude. The image above depicts a 2 second readout of low and high amplitude gamma.

Researchers need more detailed information than the consumer-strength app that comes with the Muse. So they've written sophisticated industrial-strength software that provides data suitable for research. It shows brain-wave frequencies, intensities, and time. The time stamps reveal how quickly brain function is changing in response to stimuli. Turn to the color insert near the center of the book to see the full-color versions of the images.

Here's what a typical Muse research readout looks like. From left to right, you see brain-wave frequencies. On the far left is delta and on the far right is gamma, with all the others in between. The dotted white lines running left to right represent 10 seconds of recording.

Look at the color versions in the center insert.

4.12. Muse readout of someone in a calm state of ordinary consciousness.

There are five shades of color on the readout. Dark blue represents the least amount of activity, while the red represents the most. So a brain that's very active in anxious beta, for instance, would show the red in the 20 Hz range. A brain active in healing theta would be red in the 6 Hz range. Green represents average activity, yellow increased activity, and light blue decreased activity.

This person is in an ordinary state of consciousness, just like you and me when we're awake, relaxed, and calm. Brain function is normal, so most of the frequencies are in the green. Not too much brain activity, not too little. The time stamp shows that, in each 10-second period, nothing much changes.

FIVE SECONDS TO BLISS

I don't want to throw an overwhelming amount of information at you, but this really is simple.

- Left to right: frequency
- Top to bottom: time
- Color: intensity

Got it? Now let's take a look at some other Muse readouts. The person who had normal brain function now sits down, gets comfortable, closes her eyes, and starts her EcoMeditation practice.

Look at the color versions in the center insert.

4.13. Going into EcoMeditation.

Wow! Look at the middle of 4.13. Her brain function changes radically. Within less than 5 seconds, she's producing more delta and theta waves, the signature waves of intuition, healing, and connection with the universe. That's the oval-shaped flare on the left, in the frequencies of 0 to 8 Hz.

Notice that there is one oval flare as she starts meditation and a second just after she goes into it fully. This "double bubble" is a pattern I've observed in dozens of recordings.

Right after the first bubble, you see a steep drop in all the high brain-wave frequencies, as selfing turns off. That's the PFC shutting down. It's abrupt and, to the meditator, "it feels like falling off a cliff," in the words of renowned neuroscientist Andrew Newberg. The 40% drop in PFC function measured in meditators kicks in and you're suddenly in a radically altered state of consciousness.

The "self" you're usually so preoccupied with just goes away. EcoMeditation pulls the plug and the Me Show we discussed in Chapter 2 goes dark. You're one with the universe. In Paul Brunton's words, "The outer world vanishes utterly."

But then you have a second oval flare of delta and theta, and after that perhaps a third and fourth. Your connection with the universe kicks in. You're surrendering to the implicate order of creation, no longer selfing. Brunton describes it as: "The sensation of being enclosed all round by a greater presence, at once protective and benevolent."

The anxiety frequencies of beta, in the 20 Hz band, go dormant. The whole brain is quiet. There's a normal amount of alpha, especially low alpha around 8 Hz. That's the bridge between the conscious mind's frequencies and those of the subconscious and unconscious minds. Your whole brain is functioning in an integrated manner. Image 4.14 shows what your Muse readout looks like during that period.

You have normal levels of theta and delta, with occasional flares of increased activity. While your conscious mind and anxiety frequencies are suppressed, your intuitive and connection capabilities are fully online. Imagine staying in this state of Bliss Brain for a while. Ten minutes. Thirty minutes. An hour. It feels amazing! "In deepest contemplation," writes Brunton, "both egolessness and blissful peace can be experienced."

4.14. Deep in EcoMeditation.

But eventually an alert chimes and your day begins. The middle of image 4.15 shows what happens to the brain when someone comes out of EcoMeditation.

4.15. Coming out of EcoMeditation.

Notice that there's another "double bubble" here, fainter than the one experienced going into EcoMeditation. The brain is preparing itself to return to a normal state, integrating information from Bliss Brain into everyday reality. When the meditator opens her eyes, her brain returns quickly to its normal state.

4.16. After EcoMeditation.

Yet it's not the same normal that she was in when she started meditation. Andrew Newberg observes that, "The brain that comes back from the enlightenment experience is not the brain that entered it."

Image 4.17 shows these two recordings side by side. The first is the "normal" she experienced before she started her EcoMeditation practice. The second is her brain's normal after 30 minutes of meditation. You'll see it's noticeably calmer and more integrated. There are little blips of dark blue even in the midst of the green.

That's because she's experiencing the aftereffects of her meditation in her daily life. Her brain isn't working as hard, because it's running more efficiently. Like a car that just received a tune-up, it's sucking less gas even as the engine runs better. She's built an hour of resilience, the capacity to carry the traits of Bliss Brain into everyday experience. She's happier, more peaceful, more creative, and better able to handle stress for 24 to 48 hours. In Chapter 7, we'll review practical examples of how this builds resilience.

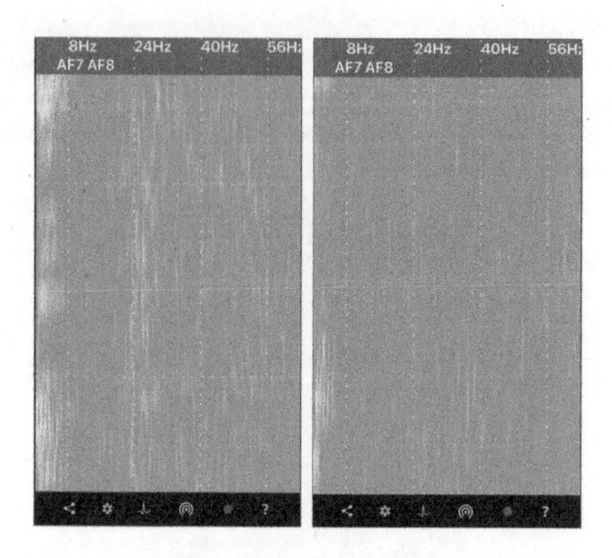

4.17. Before and after, side by side.

These recordings were made of an experienced meditator with around 3,000 hours of lifetime practice and around 7 years of using EcoMeditation. Yet I've seen these same patterns in people attending a weekend retreat after just a few hours' practice. As you use the techniques in this book, you're likely to feel subjective changes immediately. Neurologically, you'll be mimicking the way the brain of a One Percenter fires and wires.

HOW ECOMEDITATION CHANGES BRAIN FUNCTION

My colleague Peta Stapleton, PhD, runs the masters psychology program at Bond University. She's one of the most innovative researchers in energy therapies. Peta and neuroscientist Oliver Baumann, PhD, performed a randomized controlled trial of 24 people using EcoMeditation for the very first time.

After randomization, subjects were tested using a high-resolution MRI machine. After the first scan, they were given a 22-minute audio track and instructed to listen to it once a day. One group got EcoMeditation. The audio for those in the control group had them perform a mindful breathing exercise while recalling a recent vacation. The audio components were designed to be as similar as possible. This controlled for background music, voice, mindfulness, and breathing—everything except for the seven steps of EcoMeditation itself.

After listening to their tracks every day for 4 weeks, participants came back to the lab for follow-up tests. There was little change in brain function in the control group. But in the EcoMeditation group, we found significant differences in two important brain networks.

The first was an increase in connectivity between the right hippocampus and the insula. The hippocampus, seat of emotion, learning, and memory, showed increased connectivity with the right insula, central to kindness and compassion. Secondly, we found reduced activity in the medial prefrontal cortex, the "seat of the self" and also one of the two regions making up the DMN.

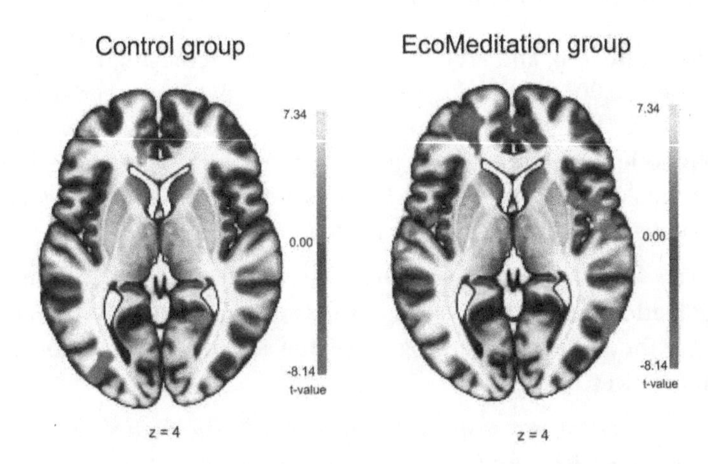

4.18. Changes in connectivity in the two groups. Blue through purple indicates lessened activity while red through yellow shows heightened activity. The blue area in the front of the brain at the right is the medial prefrontal cortex, one of the two poles of the DMN. The bright red and yellow spots on the right are the insula. See full-color version in center of book.

These findings suggest that compassion increased in those practicing EcoMeditation, and that there was strong positive emotion associated with it. This includes both compassion for self as well as compassion for others, since EcoMeditation has you send kind intentions both ways. At the same time, participants benefited from a shutdown of one of the two poles of the DMN, freeing them up from its constant self-absorbed chatter.

Though biological changes between the two groups were marked, changes in their psychological characteristics were more nuanced. Both groups more than doubled their score on a scale of mystical experiences, but anxiety and depression increased significantly more in the EcoMeditation group. This may have occurred because the audio track focused them on compassion, which can put you more in touch with the suffering in the world, as suggested by their highly activated insulas.

A side effect was marked reduction in the activity of the left dorsolateral prefrontal cortex (DFC). Other meditation studies show the DFC working hard to suppress the DMN. But in EcoMeditation, the left side or "logic lobe" of the DFC relaxed as the DMN quieted down, and participants went into a "flow" state.

THE TEN-THOUSAND-HOUR RULE?

There's no long and arduous time frame for achieving these elevated states. No "10,000-hour rule." You don't even need to spend 100 hours before you see benefits. Those in the Bond University MRI study produced substantial brain function changes in just 5 hours spread over 14 days. Because EcoMeditation is science based, sending real-time signals to your body, you're likely to feel different the very first time you try it.

That's why I'm so passionate about sharing these techniques with you. EcoMeditation brings the benefits of the One Percenters to novices immediately, without the huge learning curve that accompanies traditional meditation practices.

I can't overstate my respect for the pioneers who figured out the principles of meditation and brought them to humanity, nor can I overstate my excitement about the way that modern scientific understanding is turbocharging their insights.

Some who've followed the long and arduous path to bliss that I followed, using conventional practices, sneer at EcoMeditation as a shortcut. Could all the laborious effort they put in have been a waste of time?

Esalen Institute has been teaching advanced therapies and philosophies since the 1960s. It is regarded by many as the birthplace of the human potential movement. When I first proposed a retreat to the program manager there, and described EcoMeditation, she scoffed.

"We teach every major meditation tradition here," she emailed back to me. "I can assure you based on our long history that meditation cannot be that quick or simple."

But could it be?

Eventually, Esalen relented and I offered a weekend EcoMeditation retreat there. We had a research team on hand to perform a pilot study, and when the data came back from the lab, we were amazed at the results.

We measured cortisol, the master hormone of stress. We also measured secretory immunoglobulin A, SIgA for short. SIgA is a Y-shaped antibody molecule found in body fluids like saliva, mucus, tears, and sweat. It neutralizes invading viruses. The prongs of the Y stick to the "spikes" of several types of coronavirus. The virus uses those spikes to attach to cells and infect them.

But once a SIgA molecule has adhered to a coronavirus spike, it's deactivated. SIgA then signals specialized cells which destroy the virus. Levels of SIgA are one measure of the strength of your immune system.

After the weekend, the baseline cortisol levels of the 34 participants dropped by 29%. Their levels of SIgA rose by 27%, though the results were not statistically significant because of the small number of participants. Their resting heart rate, a measure of overall health, dropped by 5%, while their pain levels decreased by 43%.

We also measured their psychological well-being, and found that anxiety and depression declined by over 25%. Happiness went up by 11%. All these remarkable benefits come from just a single weekend of practice.

It's Never Too Late

by Vinnie Vincenzo

I've led a rough life. The drug trade is hard—I'm one of the only people from my past who isn't dead or in prison. I've hurt people, and people have hurt me. I've had a gun pointed at my head more times than I can count, and I've pointed guns at other people. I'm not proud of what I've done.

I joined Narcotics Anonymous 18 years ago. But I didn't believe in that stuff, I just went there to hide. I had a cast on my right leg, all the way up to my hip, and I couldn't get around. I pretended to be wanting to get clean, just so I could get away from the people who were trying to kill me. I went into the NA halfway house. That was a place none of my homies would be looking for me.

I slept on a concrete floor in a basement for 5 months. They have a group therapy circle and I start going. I don't feel anything. I just pretend like I feel something, so they won't kick me out. When I close my eyes, all I can see is the people I've pointed a gun at. I can't stand it. Most of the guys in the joint are pretty hard. I don't think they're going to make it. They'll be back in prison, or dead, soon enough.

One day a little girl, 5 years old, the daughter of the landlord, comes down to the basement. She says she's going to pray for me. She's not allowed to be there, but there she is. She says, "God, help Vinnie's leg. Help his heart. Help him feel your love. Amen."

She goes away and I start crying. Real waterworks stuff. I feel like my heart is cracking open. My whole body shakes. I can't stop it, I can't stop the tears, it feels like I'm washing out a lifetime of pain. I feel the presence of a something, a something that's protecting me even though I've done so many bad things.

After that, all kinds of things start to change for me. I've made a lot of money, and I've given a lot of money away. So I'm invited to be part of the state governor's task force for narcotics rehabilitation. I get to know the governor, and I talk about how people can change. I don't see dead people anymore when I close my eyes.

Now I'm in this tapping workshop with Dawson. We're doing EcoMeditation each day. The second time I do it, I close my eyes, I see these three figures around me. They're beautiful white angels. They're together under a white tree. I feel their love for me. They're more real than real, just like when I was sleeping in the basement of the halfway house. I know these angels are there to guide me. Dawson tells me that the more I pay attention to them, the more they'll connect with me.

I go to the joint sometimes to talk to the guys there about how they can change. I shiver sometimes, I think how close I came to being one of those guys behind bars. Or dead. Love changed me. My angels saved me. I find them every day in EcoMeditation.

I want my life to mean something. If I can help just one con know he's valuable, it will have been worth all the pain. It's never too late.

As Vinnie discovered, EcoMeditation's simple techniques can put you in Bliss Brain quickly, just by following the seven steps. When you can

4.19. EcoMeditation retreat at Esalen Institute.

accomplish a goal that efficiently, do you really need to take the hard path? I believe that giving people an easy alternative will result in millions more meditating. I and other traditional meditators sweated through thousands of hours of trial and error—so you don't have to.

HOW DOES SCIENCE MEASURE THE MEDITATIVE STATE?

In their book *Altered Traits*, science journalist Daniel Goleman and neuroscientist Richard Davidson point out that meditation is "a catch-all for myriad varieties of contemplative practice, just as sports refers to a wide range of athletic activities. For both sports and meditation, the end results vary depending on what you actually do." Spend three sessions at the gym lifting weights each week and you'll have a very different body from the one you'd sculpt by swimming.

There are even big differences within weight lifting. Recent research demonstrates that you build muscle much more quickly when you lift weights very slowly, that rest periods are important, and that you don't need to spend more than 30 minutes a session for effective bodybuilding. This is in sharp contrast to the old model of spending hours each day in the gym "pumping iron."

The same caveat applies to scientific studies of meditation. Some scientists lump all different types of meditators together, but fMRI research shows that different meditation methods produce different effects in the brain. Noticing your thoughts and correcting mind wandering require mental effort. This shows up as activation of the PFC. Loving-kindness doesn't require the same degree of mental control and thus produces a more positive mood as well as deeper physical relaxation.

To find the ideal "dose" of meditation, it's necessary to determine exactly *when* your meditation starts. It doesn't start the moment you sit down and close your eyes. Some people sit for 10 minutes, and from a few seconds after they close their eyes, they're in Bliss Brain. They make every one of those 10 minutes count.

Another meditator might sit for 90 minutes, yet may not enter Bliss Brain in that whole period of time. At the end, they've just been sitting quietly with their eyes closed for an hour and a half. That's a good thing in and of itself, but they didn't get even the 10 minutes of meditation the first meditator got, even though they spent much more time.

How do you know when you're "in" meditation?

Science has an answer to that question.

I did a large-scale study at a workshop taught by Dr. Joe Dispenza. We had a team of neuroscientists hook up 117 people and measure their brain function before and after the workshop. We also measured change from the start to the end of particular meditations. The metric we used for entering the desired state is called "time into meditation."

4.20. Researcher Judith Pennington hooks participants up to a Mind Mirror EEG at a workshop.

We defined this as *the ability to maintain a stable alpha brain-wave state for at least 15 seconds*. Once participants were able to do this, we considered them to be "in" meditation.

SUSTAINED ALPHA

When you close your eyes, you immediately produce more alpha. As you start to focus on meditating, you produce still more. But the depth of your entry into alpha varies as you cycle in and out. The person who sits for 90 minutes with her eyes closed and doesn't actually meditate has a restless brain controlled by the DMN. It keeps introducing worried thoughts, pulling her out of alpha. She can't maintain alpha for as long as 15 seconds at a time, so according to that definition, she never actually meditates, despite her extended time commitment.

Though it's common to consider people who sit in eyes-closed silence regularly and for long periods of time as dedicated meditators, the truth is more complex. Hooking people up to EEGs has shown scientists that people who are doing this may actually be spending little or no time in sustained alpha, while some "quickie" meditators are deep in alpha and getting the full benefits of every moment they spend in that state.

GAMMA

When you sustain alpha brain waves, several other waves change. High-frequency beta waves, "the signature waves of stress," shrink, decreasing in amplitude. Low-frequency waves, theta and delta, grow, increasing in amplitude.

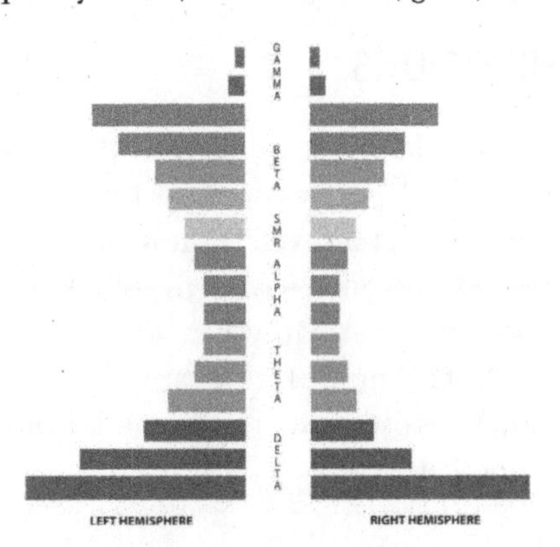

4.21. Gamma are the waves in red at the very top of the screen. See full-color version in center of book.

When these slow waves increase, the fastest wave, gamma, is produced by the brain. Gamma is the signature wave of the "flow" state and represents the *synchronization of information from many different brain regions*. It's usually found in highly creative people, as well as ordinary people having a moment of insight. It's also observed in states of mystical union. When studying advanced yogis, Davidson found that their brains had 25 times the gamma activity of ordinary people.

Gamma is also associated with feelings of love and compassion, increased perceptual organization, associative learning, synaptic efficiency, healing, attention, and states of transcendent bliss. Neuroimaging studies have shown that gamma waves synchronize the four lobes of the brain across frequencies and engender *whole-brain coherence*.

That's a long list of benefits, all hallmarks of the high-performance brain. And while the studies of Tibetan monks show that big gamma appears only in adepts, EcoMeditation research shows it appearing even in novices.

Neurofeedback expert Judith Pennington, studying eight subjects, wrote that: "EcoMeditation produced extraordinarily high levels of Gamma Synchrony . . . . During EcoMeditation all participants generated high-amplitude 45–65 Hz gamma frequencies that resulted in high-value Gamma Synchrony patterns, some of which were at the top of the statistical range found in the entire Mind Mirror database."

Some of these were people who'd never meditated before, and one of them achieved higher scores than anyone else in the entire database of over 10,000 Mind Mirror scans.

TOTAL LIFETIME HOURS

When we read stories about a monk who has meditated for 10,000 "total lifetime hours" or TLH, it's easy to be impressed by that statistic. What is even more important, however, is the *quality* of that meditation. If, like the Tibetans in Davidson's and Newberg's studies, they've been trained by experts, are steeped in ancient lineages, do long retreats, and live in monasteries, that 10,000 TLH number is meaningful. For Westerners without this structure, much less so. Just spending a lot of time at something doesn't mean you're good at it.

4.22. The traditional training for Tibetan monks can produce advanced brain states.

One Foot on the Gas, One Foot on the Brake

Steve Cheales was a close friend of mine in Brooklyn, where he was born and raised. He was close to his dad, Samuel, and when I was visiting Steve, Samuel was frequently there too, so I got to know him well. Elderly, gray-haired, and permanently hunched over, he had a New Yorker's bright wit and dim view of the world.

Steve didn't own a car, while Samuel had a big old Buick Skylark. When the three of us were going places, Samuel drove us. I would sit in the front passenger seat on account of my super-length legs.

Being a passenger in Samuel's car was nerve-wracking. He would hit the gas, throwing us back into our seats with a jackrabbit start. Then he'd hit the brake, throwing us forward. He'd then speed up again, after which he'd repeat hitting the brake. The entire journey was accomplished in this stop-start manner. Driving with Samuel was like being a golf ball that was rattling around in a tin can.

The reason Samuel could switch between stop and go so quickly was that he drove with one foot on the gas and the other on the brake. I watched his feet. His right foot would pump the gas, speeding us up. Then his left foot twitched, hitting the brake. Right, left, right, left. Staring blankly at the road ahead, he was completely unconscious of this habit.

Samuel had learned to drive when he was 18 years old. He'd been driving for over 40 years. But he was still a terrible driver. Time had not improved him. Just the opposite: It had reinforced his behaviors through neural plasticity till they were engraved in his basal ganglia.

Simply spending a large amount of time in an activity can make you very good at it—the famous 10,000-hour rule. In his book *Outliers: The Story of Success*, science journalist Malcolm Gladwell talks about how extended repetition produces expertise. Whether you're a chef, a guitarist, a mathematician, a realtor, or a surgeon, experience counts. Accumulating 10,000 hours takes a massive time commitment; think 20 hours a week for 10 years.

But like Samuel, *doing the same thing badly for a long time* doesn't make you better at all. It just reinforces bad habits. That's why athletes hire coaches to improve their form. They don't want to just repeat a technique; they want to repeat it correctly, training their bodies to the point where the perfect form becomes a reflex.

Don't Fuck with My Serenity

Herman was tiptoeing around the apartment he shared with Jill. Jill was meditating, and Herman didn't want to disturb her. It was just 6 months since Herman had moved in with Jill, a longtime meditator, and while he didn't meditate himself, Herman was a peaceful guy who wanted to give Jill her space.

That particular morning, Herman was unsuccessful. He made one tiny sound too many. Jill opened her eyes to glare at him angrily and screamed, "Don't fuck with my serenity!"

Jill, even though she sat down each morning and closed her eyes, was not truly meditating. Her boiling emotions were right near the surface, ready to reach out and impale people close by. She was reactive to the outside world, in Caveman Brain, even though to an observer it might have appeared that she was a faithful and consistent meditator. She hadn't made the subject-object shift, disidentifying with her local self and becoming absorbed into Bliss Brain.

That's the fate of many longtime meditators. During the peer review of this book, one of my colleagues was very skeptical about the title. He said, "Dawson, I spent a year in an ashram when I was 23 years old. We meditated every day. It was miserable. There wasn't even pleasure there, certainly not bliss."

Unless you're in the Awakened Mind state, with the brain-wave patterns characteristic of the mystics, you can sit for years with your eyes closed and achieve little more than expertise in lighting up your DMN.

That's the pickle the inefficient meditator finds herself in. All those hours, thousands of them, and she's simply been reinforcing the neural wiring that has one foot on the gas, the other on the brake. Just like hiring an athletic coach, it's vital to build your TLH in a type of meditation that's predictably putting you into a sustained alpha state.

That's why EcoMeditation has you picture a big empty space between your eyes. This simple visualization puts people in alpha immediately. If you use an audio track to practice EcoMeditation, you'll hear the empty space instruction over and over again. That keeps you in alpha for the entire time you're sitting.

THE LIFE VISION RETREAT

Practicing this way regularly catalyzes changes in your brain. Doing it while on an intensive retreat, more so.

My life and work changed significantly after my first long retreat at age 30. A wonderful new career in book publishing came to me through a series of synchronicities. I married and had two extraordinary children. I was inspired to write my first book. I moved from the concrete canyons of New York to the redwood-covered mountains of Northern California.

A retreat involves stepping out of the rat race. The act of abandoning the familiar traces of your routine, erasing the cues of your everyday life, gives you fresh perspective. You remove yourself from all the environments that keep you thinking the same thoughts, believing the same beliefs, and creating the same creations. These same old sights and sounds keep you re-creating each day's new you in the pattern of the old you. But go on a retreat, take yourself out of the familiar, and you open up the possibility of transformation.

Empowered by the changes that reflection produces, I began going on New Year's retreats every year. When other people asked to join me, I began structuring the experience for a group. Each year we go away to someplace gorgeous, like Sedona, Hawaii, Costa Rica, or San Diego, for 7 days.

4.23. Life Vision Retreat participants.

We spend the first 3 days releasing the old. We practice EcoMeditation daily. At first it takes participants about 4 minutes to induce Bliss Brain, but by the third day they're usually there in about 90 seconds. We then open our awareness to guidance, asking, "What's the universe's highest possible vision for my life?" By the end of our 7 days together in Bliss Brain, we each have an inspired written vision for the year ahead.

The retreat where I turned 60 was special. I imagined the universe saying, "What do you really want this coming year?" At the top of my list was to be able to enter Bliss Brain quickly. Not to sit for 20 to 40 minutes, the length of time it usually takes people to drop into "the zone." Instead, to be able to close my eyes and be there instantly.

Over the next few months, my "time into meditation," acquiring a stable alpha state, began to drop. To 20 minutes, to 15 minutes, to 12 minutes. On auspicious days, to under 5 minutes. When Andrew Newberg tested experienced meditators, he found they could enter that state as soon as they closed their eyes.

Richard Davidson's Tibetan monks went one better. He found that when they were being given instructions *in advance of* an experiment, their brain waves changed to a meditative state. The *mere anticipation of meditating* in the near future produced a change. Those who'd spent more of their TLH in retreats did better than those who'd meditated alone.

You might not be there for a while, but with practice, your "time into meditation" and acquiring a stable alpha state can drop from the best part of an hour to under 10 minutes. Those who train themselves deliberately at an EcoMeditation retreat often achieve Bliss Brain in under a minute.

STRENGTH IN NUMBERS

I've pored through the neuroscience literature to find out what really moves the needle. *Consistent* practice, *retreats*, and *intensity* are all associated with greater brain activation.

To find out just how quickly EcoMeditation could shift people, I studied 208 participants at a very brief retreat. This one lasted just a single day. I explained each component of the practice to the precious people there, plus the physical changes that meditation produces. We meditated together in the morning and afternoon. We engaged in a lively Q&A session, then ended with a closing circle.

My colleagues and I were surprised at how much people changed in just 6 hours. Their anxiety decreased by 23% and their pain by 19%. Their happiness increased 9%. When we followed up with them 6 months later, we found that they'd maintained their gains in anxiety and depression levels over time, and were even happier. That's a big payoff for 6 hours of group practice. Whenever you have the opportunity to *practice with others*, grab it with both hands.

A challenge when reading meditation studies is that there are so many different kinds of meditation. EcoMeditation focuses on physical cues, but other practices trigger a contemplative state in other ways.

SEVEN TYPES OF MEDITATION

A study at Germany's famed Max Planck Institute in Leipzig, Germany, differentiated meditation into three styles. The first style focuses on the **breath**. The second emphasizes **observing one's thoughts**. The third evokes **feelings of loving-kindness**. By differentiating meditation into these three common forms, the Planck researchers were able to analyze the brain patterns typical of each. They are different, as we will see in Chapter 6.

To the three Planck schools I would add four more: Those that use **movement**; those that use verbal, vocal, or other **auditory cues**; those that employ **visualization**; and those that rely on **imitation** or inspiration from others.

Sufi dancing, tai chi, and Buddhist walking meditation involve physical movement. Conscious deliberative movement can raise your level of self-awareness. Other meditative styles that use physical awareness include Yoga Nidra, body scanning, and progressive muscle relaxation.

Verbal or vocal cues include chanting, saying a mantra, or reciting a prayer. Other auditory cues that can induce meditative states include gongs, hymns, reciting the names of God, the sound "Om," and singing bowls.

Visualizers can picture a journey through their chakras or focus on the image of a saint. They can visualize a scene of deep peace, such as a special spot in nature, or a sacred image such as a yantra.

4.24. Sufi dancing.

Meditation styles that draw inspiration from others may involve reading a sacred text. The writing of a saint can lead you to inhabiting the enlightened state. Renowned theologian Huston Smith, author of *The World's Religions*, began each day's meditation by reading from one of his favorite teachers. Imitating an adept's mindset gives the meditator a template for enlightenment.

4.25. Tibetan yantra. See full-color version in center of book.

Some breathing meditations involve complicated breathing instructions such as closing one nostril, then the other. Or breathing very fast for a while, then very slowly. Others have you simply notice your breath. Eco-Meditation has you slow your respiration down to 6 seconds per inbreath and 6 seconds per outbreath, which induces heart coherence.

COMPASSION RULES

A key Planck study compared people experiencing *empathy* with those actively engaging *loving-kindness*. It found that after just 8 hours of practice, completely different brain regions were activated. A second study showed that after a similar training period, "strong echoes" of the brain patterns of experienced meditators were found in novices.

Research shows that loving-kindness and compassion produce more beneficial brain changes than other types of meditation, and we'll explore the profound implications of this in Chapter 8.

A study comparing a non-meditating group with groups doing mindfulness and loving-kindness meditation found marked differences between the three. The researchers measured telomeres, the "end caps" of chromosomes. These are regarded as the most reliable biological marker of aging, since they get shorter by about 1% a year.

Those in the loving-kindness group "showed no significant telomere shortening over time." Those in the mindfulness group fell between the compassion group and the no-meditation group. By way of contrast, a study of the stress hormone adrenaline linked it to "drastically reduced lifespan." Caveman Brain kills, while Bliss Brain heals.

The Universe and Back

My wife, Christine, was reading a bedtime story to her granddaughters, Natalie (6) and Kiera (3). The book was called *Guess How Much I Love You* by Sam McBratney.

In it, Little Nutbrown Hare is trying to tell Big Nutbrown Hare how much he loves him. He opens his arms, but that isn't enough. He stretches to bigger dimensions, but they're still not enough.

Finally, just before he falls asleep, he finds the scale he's been looking for. He tells Big Nutbrown Hare, "I love you right up to the moon."

As Christine cuddled the girls in bed later, Natalie said, "Grandma, I love you right up to the moon!" Christine responded, "I love you right up to the moon!"

To which Kiera responded: "I love everyone—to the universe and back!"

Kiera had never read the studies showing the efficacy of loving-kindness and the power of extending it to the whole universe. Yet at just 3 years old, her experience of love extended to the furthest boundaries of time and space.

With luck and practice, all of us can get back to the primal experience of connection with the whole universe we had when we were 3.

I had a conversation with psychiatrist Dan Siegel, MD, shortly after the publication of his book *Aware: The Science and Practice of Presence*. We both amended the instructions for our meditation methods to emphasize compassion after being convinced by these research findings. Buddhist traditions have found these three conceptual steps helpful:

First, "May *I* be filled with loving-kindness."

Then, "May *you* be filled with loving-kindness."

Finally, "May *all* be filled with loving-kindness." You'll find this chant in the Deepening Practices section at the end of this chapter.

The first step, focusing on the self, makes the experience relevant and concrete. Second, you extend love to someone you know personally. From there, you widen the scope of your compassion to the infinity of the universe. That creates some of the most dramatic changes in brain function.

HOW LONG TO START EXPERIENCING BENEFITS?

You don't need to be a Tibetan monk with 30,000 TLH to experience the benefits of meditation. Early studies showed that 8 weeks of mindfulness meditation had an effect, and later ones showed that even 4 weeks made a difference.

A startling study performed by investigators at the University of North Carolina showed that *just 4 days* of meditation were associated with increased cognitive flexibility, creativity, memory, and attention. The lead researcher said that these "profound improvements" were "really surprising" and that they are "comparable to the results that have been documented in far more extensive training."

4.26. Young children can have rich spiritual lives, even though they may lack the vocabulary to describe them.

A study of beginning meditators found that their practice became progressively more enjoyable and easier during 10 weeks of meditation. This held true whether they used a loving-kindness style meditation, observed the flow of their thoughts, or focused on the breath.

Keep it up for months—then years—and the very structure of the brain changes fundamentally. You look the same on the outside, but inside your skull the wiring is very different, as we'll see in Chapter 6. In Chapter 5, we'll show how the experience floods your brain with pleasure hormones. Your stress levels drop and you become exceptionally resilient, outlined in Chapter 7. From the first meditation to the 10,000 TLH level, your brain continues to evolve.

Goleman and Davidson divide the length of meditation practice into three levels.

Beginners experience benefits quickly. Regulation of the amygdala and DMN is found in novice meditators after just 30 hours of practice spread across 8 weeks. *Compassion meditation produces bigger changes*; in just 7 total hours in 2 weeks, increased connectivity was found in the brain's circuits for positive emotions and empathy. Even *7 minutes* of loving-kindness can increase positive mood and social connection.

They define **long-term meditators** as those with 1,000 to 10,000 TLH. This indicates sustained daily meditation and perhaps retreats or meditation courses. These meditators can sustain their attention, and the mind wandering typical of the DMN decreases. Brain circuits for emotional regulation are enhanced, and stress hormones like cortisol decrease.

Adepts are those with 10,000-plus TLH. For these One Percenters, meditation is effortless. Their attentional control is complete. Their hearts and brains are bonded in resonance. Brain function and structure continue to evolve, and the brain states found in meditation pervade their daily lives. The brain's default state now becomes the meditative mode.

SYNCHRONICITY, PRECOGNITION, AND OUT-OF-BODY EXPERIENCES

Adepts routinely experience anomalous states of consciousness. A survey of 1,120 meditators found that there were statistically significant correlations between the length of time people had been meditating, and 43 out of a list of 50 "anomalous experiences."

These included phenomena such as synchronicity, clairvoyance, telepathy, out-of-body experiences, psychokinesis, and precognition. These experiences are described more fully in my book *Mind to Matter*, and it's fascinating to me that the nonlocal experience of Bliss Brain is associated with such a big variety and frequency of such "psi" phenomena.

HOW LONG A MEDITATION IS BEST?

There's a concept in medicine called the minimum effective dose, or MED. For instance, consider the time-honored advice to take two aspirins for a headache. If you take half an aspirin, it's probably not enough to take your headache away. Two aspirins is the MED.

So if 2 aspirins take a headache away in 30 minutes, will 20 aspirins take the headache away in 3 minutes? No. The MED does the trick and more than that is pointless or even dangerous.

With meditation, it's important to find the MED that's right for you.

So what's the perfect duration of your meditation session?

4.27. There's a minimum effective dose for medication—and for meditation.

Here the scientific evidence is ambiguous, but most studies suggest time frames of *over 25 minutes*. Andrew Newberg found that it takes 40 to 60 minutes to achieve a deep contemplative state, as the parietal lobe that orients you to the outside world shuts down. That was the time frame between starting meditation and feeling oneness with the object of contemplation. He found that even advanced meditators could not still their minds for a full 60 seconds; novices could not last 10 seconds.

A research group at Harvard University examined the effect of 27-minute guided mindfulness sessions over the course of 8 weeks. They found increases in the gray matter in regions of the brain related to emotional regulation, learning, memory, perspective taking, and sense of self.

Yet in just a few minutes, observable changes take place. After just the first 8 minutes of mindfulness, attentional focus improves. If you're a novice, start with 10 minutes. If experienced, try 30 to 45 minutes. When you have uninterrupted leisure time, 90 minutes of Bliss Brain is delicious. Most days I clear the first 60 minutes of my day for meditation.

INTENSITY

The *intensity* with which meditation is practiced also makes a difference. Among Tibetan monks, the most pervasive brain changes were found in those who went on *long, intensive retreats*. While consistent daily practice is good, the extended spurts of meditation you do in retreats are better.

Newberg found that when you intensify the contemplative experience, the parietal lobe shuts down and selfing ceases in less than 10 minutes, not the 40-plus minutes usually required. In one case he studied, this occurred in 6 minutes.

For novice meditators, my recommendation is to start small. I'd rather see you devote 10 minutes a day to meditation and be consistent than to pledge allegiance to a 60-minute practice. Starting small makes it more likely that you'll stick to your goal.

I have many free meditations on Insight Timer, the world's biggest meditation app. You can find them through the link at the end of this chapter. They're mostly under 15 minutes. I recommend a consistent short meditation over an intermittent long one. You'll understand the biochemistry of *why* after reading Chapter 5.

SUSTAINING BLISS BRAIN

How do you sustain Bliss Brain during a meditation once you've achieved it? I searched the scientific literature for the answer to this question and found little of help. So I took the question into meditation.

Once those gamma flashes began in my brain, I got the answer: *Make your periods of mind wandering as brief as possible.* The moment you observe your mind wandering, move it immediately back into Bliss Brain. You'll find an exercise on how to do this in the Deepening Practices at the end of this chapter.

Perhaps you have 1-minute periods in Bliss Brain, after which your mind wanders for 5 minutes. Catch yourself a little earlier and you can reduce your mind wandering period to 4 minutes. With practice, to 3 minutes, then less. I think of this as the Spiral of Practice.

When your mind wanders, bring it back to focus immediately. Spend as little time in the DMN state as you can. Spend as much of the meditation as possible in the alpha state.

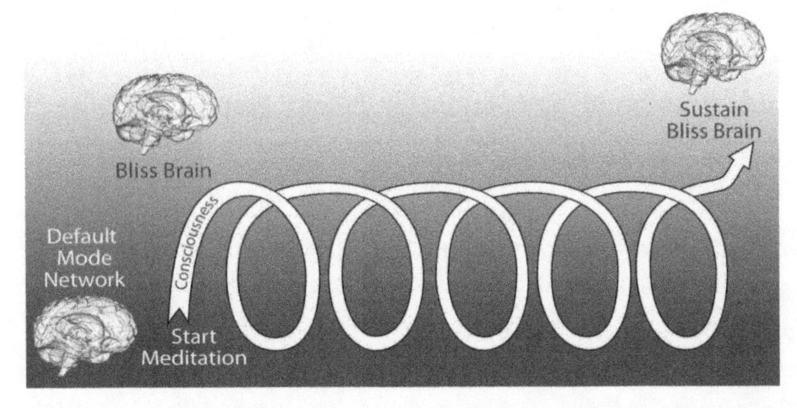

4.28. The Spiral of Practice. We start meditation and after a while enter Bliss Brain. Then the Default Mode Network kicks in, our minds wander, and we fall out again. We come to awareness, shift ourselves back in, and reenter Bliss Brain. Once we repeat the cycle often enough, we can fully integrate with Bliss Brain.

Make your meditations short but regular. Intensify them. If you feel the physical sensations of bliss, dial them up. Feel them *as intensely as you possibly can*. Rather than a marathon, aim for an intense sprint.

This way, you'll enter Bliss Brain quickly, make most minutes of your practice effective, and unlock the cascade of pleasurable neurotransmitters we'll consider in Chapter 5.

Falling Down the Ladder

I've been friends with Andrew Vidich, PhD, for some 40 years. He's been meditating since he was 17 and is one of the most inspiring teachers I know. He's written two books. *Love Is a Secret* is about the psychological states that seekers go through on their way to ecstatic oneness with the divine.

Light upon Light describes the illumination provided by the lives and words of renowned spiritual masters from Rumi to 20th-century mystic Sant Kirpal Singh. Every time I visit New York, Andrew and I get together and spend long hours reflecting on the spiritual experience.

Over a superb cup of drip coffee at Blank Slate coffee house, I asked him, "Andrew, how can you get into the deepest meditative state quickly?"

His reply startled me. "It's not what you do in the hour of meditation that counts; it's what you do in the 23 hours you're not meditating."

"How so?" I asked.

"When you're in meditation you're at a high vibrational level. If you sustain that vibrational level, or one close to it, during the rest of your day, then when you start your next meditation, the following day, you're close to your original state. But if you descend many vibrational levels down, then you have many to ascend to get back to that peak experience again.

"It's like a ladder. The further down the ladder you fall, the lower the vibrational rung from which you start climbing back again when you close your eyes to meditate the next day. Staying high on the ladder all day long is key to entering the peak state quickly when you start the next meditation practice."

4.29. It's how far down the ladder we fall after a meditation that determines the rung on which we start the next meditation.

In *Altered Traits,* Goleman and Davidson emphasize that the true point of practice is "to transform ourselves in lasting ways day to day . . . it's not the highs along the way that matter. It's who you become." They sum up

Andrew's ladder idea with the memorable phrase: "The after is the before for the next during."

The way you conduct yourself after an elevated state is the entry point from which you start the next meditation. *Living your everyday life in the highest possible spiritual state* gives you the consciousness in which to start your meditations high on the ladder.

UNLOCKING THE SECRETS OF THE ONE PERCENTERS

Thanks to science, we can now describe the brain-wave signature of the One Percenters. We've identified the Awakened Mind profile common to all mystical experience, regardless of the spiritual tradition or religion that fosters it.

We've learned the importance of sustained alpha. We've discovered that compassion-style meditation produces greater brain change than other styles, that retreats produce better results, and that intensifying the experience leads you to Bliss Brain quicker.

Science thus provides a reliable road map to the states previously known only by a small group of adepts. Today we can show the other 99% how to experience Bliss Brain. Science has uncovered the secrets of the richest "happiests" in the world, the billionaires of enlightenment. What is this precious knowledge worth?

Lunch with Warren Buffett

Billionaire Warren Buffett, the "Sage of Omaha," is the most famous investor of the 20th century. In 1962 he bought his first shares of a struggling textile company called Berkshire Hathaway for $7.50 each.

By 1965 he owned the business. It became a multibillion-dollar holding company. Today, each share is worth over $300,000.

When Tony Robbins wrote his book *Money: Master the Game—7 Simple Steps to Financial Freedom,* he interviewed all the top investors in the world. The only one who denied him an interview was Warren Buffett.

What would you pay for an interview with Warren Buffett? To ask him questions about how to maximize your wealth?

Well, that dollar amount turns out to be quantifiable. At the annual charity auction held on eBay for the Glide Foundation, an anti-poverty nonprofit organization based in San Francisco, people have paid as much as $4,567,888 to have a single lunch with Warren Buffett. That record high was in 2019.

Back in 2007, however, the lunch went for a bargain $650,100. The two men who won the bidding that year, investors Guy Spier and Mohnish Pabrai, shared what they learned during this lunch of a lifetime.

First, they said the experience was "fantastic" and "worth every penny." Pabrai added, "I think we would have been willing to pay a lot more than that."

As for what they learned from Buffett, they summed it up as three valuable business lessons:

- Approach everything with integrity.
- Get comfortable saying no.
- Do what you love.

On integrity, Buffett asked Spier and Pabrai, "Would you rather be the greatest lover in the world yet be known as the worst, or would you rather be the worst lover and be known as the greatest? If you know how to answer that correctly, then you have the right internal yardstick."

Spier said, "Buffett was teaching us to act with the right motivation—because it's the right thing to do, not because of what people will think."

Buffett believes that highly successful people say no to most opportunities. Spier noted that, "even though he is a kind man at heart, he also has absolutely no trouble enduring the momentary unpleasantness that comes from saying no. As I realized this, I resolved to get a lot better at my own ability to say no."

On doing what you love, Buffett stated, "When you go out in the world, look for the job you would take if you didn't need the money."

Ten years later, Spier observed that the lessons he learned were "less about specific investments, and more about life."

At the end of the lunch, Buffett left an enormous tip for the waiters.

Money is important, but happiness is everything. The only reason people want money is that they believe it will buy them happiness.

What would you pay for a lifetime of Bliss Brain? What is the value of learning the secrets of the One Percenters of enlightenment?

In ages past, the price was high. You had to enter a monastery, climb a remote mountain peak, give up all worldly pleasures, and spend thousands

4.30. Guy Spier and his wife, Lory, at their $650,100 lunch with Warren Buffett.

of hours in devotion. Daniel Goleman writes that "meditation . . . had for centuries . . . been the exclusive provenance of monks and nuns." You had to renounce the world, sit at the feet of masters, observe severe penances, learn esoteric rites, and undergo elaborate initiations. The process of learning the secrets of the One Percenters took many years.

No more. Science has democratized happiness. Thanks to the discoveries we've made using HRV monitors, MRIs, and EEGs, these secrets are available to anyone. You can have the spiritual equivalent of lunch with Warren Buffet every day.

The mystical experience used to be mysterious. It was a subjective state that we could barely describe or comprehend. Now we can measure it in frequencies and amplitudes. We understand the electromagnetic signature of Bliss Brain and anyone can be trained to attain it. Ordinary people can enjoy extraordinary states.

Not only that, practicing these states conditions all 24 hours of your everyday life. Davidson found that even while sleeping, the brains of adepts were bathed in high gamma. Newberg says that enlightenment "pulls the rug out from under your normal way of processing the world" and you see everything with fresh eyes. Judith Pennington found that after learning EcoMeditation, novices "were subsequently able to carry this consciousness into daily life with eyes open."

With this new brain, we create new circumstances around us. We bring joy, creativity, and flow to our everyday lives. We escape the traps of our old minds, our threadbare habits, our self-defeating behaviors, and our childhood conditioning. Embodying the secrets of the One Percenters provides more than a happy brain. It is the foundation of a happy life.

DEEPENING PRACTICES

Here are practices you can do this week to integrate the information in this chapter into your life:

- **The Stress Continuum:** Think of stress on a continuum, with red on one end, and green on the other. Right now, where are you on this slider? Set your watch or phone to chime every hour today. Each hour, record where you are on the spectrum from red to green.
- **The Binary Practice:** During your meditations this week, notice whether you're in or out of Bliss Brain. There are only two possibilities, that's why it's binary, either in or out. *Anything not in is out!* Bring yourself back in every time you're out, even if the out feels fine. Correct yourself back into Bliss Brain immediately.
- **Intensification Practice:** During this week's meditations, notice any pleasant physical sensations you feel in your body. Intensify them. Perhaps you feel a sense of love in your heart, or a tingling behind your eyes. Double it. Enjoy the sensation. Then double it again. With practice, you'll find you can "dial up" these feelings to produce an even more intense sense of well-being.

EXTENDED PLAY RESOURCES

The Extended Play version of this chapter includes:

- Dawson Guided EcoMeditation: Intensifying the Meditative Experience
- Upcoming Life Vision Retreats
- Audio Interview with Dan Siegel, PhD, Author of *Aware*
- Loving-Kindness Chant with Julia and Tyler
- Institute for the Awakened Mind
- Insight Timer Dawson Meditations
- Daniel Goleman on Extended Gamma in Long-Term Meditators

Get the extended play resources at: BlissBrainBook.com/4.

CHAPTER 5

THE BLISS MOLECULE

THE ELECTROCHEMICAL BRAIN

The brain operates in many dimensions. One is electromagnetic. We can monitor the brain's electromagnetic activities with devices that measure energy, such as EEGs and MRIs. Another dimension of brain activity is chemical. The most famous equation in history is Einstein's $E = mc^2$. E for *energy* is on one side of the equal sign and *matter* is on the other.

Like most equations in physics, Einstein's works backward as well as forward. Matter can become energy just as energy can become matter. Burn coal (matter) and we can produce electricity (energy). Add heat (energy) to water (matter) and the water changes from liquid to steam. Physicists are designing experiments to turn photons of light (energy) into electrons (matter).

As brain waves (energy) fluctuate, molecules such as neurotransmitters and hormones (matter) are created or destroyed. When we enter Bliss Brain, brain waves change dramatically. Beta waves, the signature waves of stress, virtually disappear. Alpha, theta, and delta waves expand. The gamma waves introduced in Chapter 4 appear. We'll learn more about the characteristics of this "genius wave" in Chapter 6.

Meditation changes the brain's energy patterns, which leads to changes in the chemicals the brain uses for signaling among neurons. Bliss Brain produces *a cascade of pleasure-inducing neurotransmitters and hormones*, bathing the brain in the chemicals of ecstasy.

We've all heard of hormones like estrogen and oxytocin, and neurotransmitters like serotonin and dopamine. We know they are potent molecules, affecting our moods for better or for worse. But very few people, even neuroscientists, understand how *they work together during meditation* to produce drastically altered mental and emotional states.

Bliss Brain isn't just the electromagnetic state discussed in Chapter 4. Bliss Brain is a chemical state too, associated with a unique cocktail of hormones and neurotransmitters. This mixture is so potent that it transports the brain into ecstatic highs achievable in no other way.

In this chapter, we'll identify the individual ingredients of that cocktail—then show how you can play bartender. You'll learn how you can whip up that recipe and create this unique mixture of ecstasy-producing chemicals in your own brain anytime you choose.

Let's party!

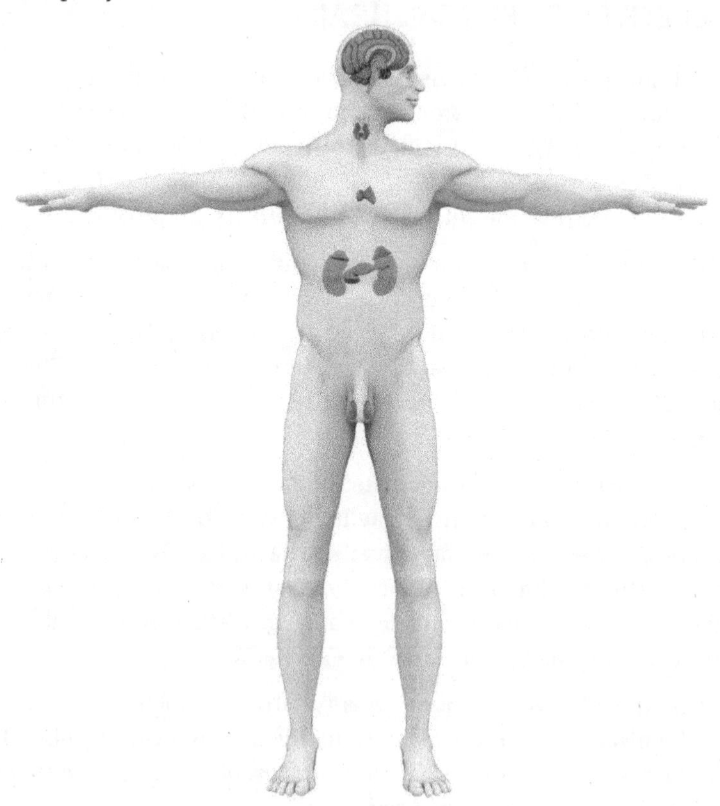

5.1. The endocrine glands.

MOLECULES OF EMOTION

Both neurotransmitters and hormones are chemical messengers, but neurotransmitters communicate *within the nervous system* while hormones chat with the *entire body.*

Hormones are produced by endocrine or ductless glands. Endocrine glands don't need ducts to siphon their hormones into the bloodstream in order to send messages. Hormones are chemically designed to diffuse quickly throughout all the body's cells. Take a dropper filled with black ink, release a drop into a glass of water, and it spreads quickly through the solution. Likewise, hormones communicate with all the body's cells within seconds.

5.2. Diffusion in seconds throughout a solution.

Hormones can act on cells far from their originating glands. Imagine your boss suddenly yelling at you, stressing you out. The adrenal glands on top of your kidneys produce cortisol and epinephrine (aka adrenaline) immediately. Within 5 seconds, your heart is pounding and your breathing becomes shallow, as stress hormones act on organs far away.

Neurotransmitters are not designed to travel those long distances. They travel only microscopic distances, between the synapses or gaps between neurons.

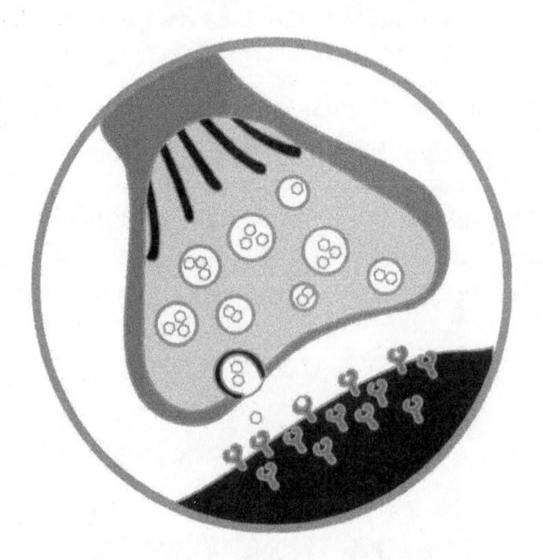

5.3. Neurotransmitter molecules in a synapse.

Neurotransmitter action is very rapid—a few milliseconds—while hormonal effects can extend from a few seconds to a few days. Both neurotransmitters and hormones, along with components of the immune system, regulate the fight-or-flight response.

5.4. A molecular key docking with a lock.

The most commonly known neurotransmitters are serotonin, dopamine, norepinephrine, acetylcholine, glutamate, and GABA. Each type of neurotransmitter fits into its own receptor on the target neuron, like a key into a lock.

Some neurotransmitters are *excitatory.* They *increase* the activity of the neuron down the line. Other neurotransmitters are *inhibitory,* which means that they *decrease* the firing of the following neuron. Glutamate is the most excitatory neurotransmitter and GABA ("the brain's own Valium") is the most inhibitory.

Though they can have dramatic effects, neurotransmitters are made from simple molecular building blocks. These are present in large quantities and require very few steps for the body to manufacture. This makes them readily available to the nervous system at any time, and they can be synthesized in seconds.

Feminist Neurotransmitters

In 1952, 25-year-old biochemist Betty Twarog made a startling discovery in a Harvard laboratory. She had no idea that it would alter the course of medicine and serve as "the cornerstone of the antidepressant revolution."

Twarog was working to solve one of several scientific puzzles laid out by Ivan Pavlov (of salivating dog fame) in his Nobel Prize acceptance speech in 1904. He speculated about what might cause muscle contractions in the digestive tract.

Twarog's discovery was the neurotransmitter serotonin. At the time, scientific belief held that there were only two neurotransmitters in the human body: acetylcholine and epinephrine.

Abbott Pharmaceuticals had recently synthesized a molecule called 5-hydroxytryptamine. They mailed samples to a range of scientists, including Twarog's mentor, John Welsh, hoping the scientists could find a use for it.

Serotonin, the natural form of 5-hydroxytryptamine, had been found elsewhere in the body earlier. It had been identified in the human gut in 1935 and in the bloodstream in 1948. The name "serotonin" is a combination of the Latin *serum* ("watery fluid"), the word "tonic" (medicine), and the chemical suffix "in."

Following a hunch, Twarog looked for serotonin in the human brain. This ran counter to current scientific thinking, which held that humans were different from other mammals, and the brain worked independent of the body. Not shy about her opinions, Twarog labeled these notions "sheer intellectual idiocy."

She discovered that serotonin functioned as a neurotransmitter in humans, and 3 years later identified it in the brains of monkeys, rats, and dogs.

The male-dominated scientific community was delighted. Not!

Twarog submitted her first research paper to the *Journal of Cell Physiology*. Not only did the esteemed journal fail to send it out for peer review; they never even acknowledged receipt of the study. When Welsh ran into the journal's editor at a conference, he told Twarog's mentor that there was no way he would ask his prominent reviewers to look at a paper by "an unknown girl."

In 1954, biochemist D. Wayne Woolley, laboring under none of these constraints, published a paper noting the chemical similarities between LSD and serotonin. He reasoned that the brain processes the natural and artificial molecules in a similar way.

After the publication of this and other papers by male scientists, Twarog's work was finally published, gaining some of the recognition it deserved.

Yet it had unintended consequences. It led to the widespread belief that mental health problems like depression and anxiety were due to chemical imbalances in the brain, and that curing them only required supplying the missing ingredients.

Antidepressants are called selective serotonin reuptake inhibitors (SSRIs). Google Dictionary tells us that this designation is "for a class of antidepressants that work by increasing levels of serotonin in the brain." Driven by the chemical imbalance concept, and fueled by the commercial success of the serotonin-enhancing drug Prozac, a massive drugs industry was born.

In rich countries, sales of antipsychotics and antidepressants have risen around fiftyfold since 1985. Prescribing practices have spread to children and elders. The number of elderly Americans taking three or more psychiatric drugs has doubled since the early 2000s. The industry generates over $14 billion a year for manufacturers.

> For most patients, studies show little benefit but plenty of side effects. These include loss of sexual function, apathy, insomnia, mania, and suicidal impulses.

Research has never proven the chemical imbalance theory, or the corresponding notion that supplying an exogenous drug can produce mental health. Ronald Pies, MD, editor in chief of *Psychiatric Times*, declared that, "The legend of the 'chemical imbalance' should be consigned to the dustbin of ill-informed and malicious caricatures."

Psychotherapist Gary Greenberg, author of *Manufacturing Depression: The Secret History of a Modern Disease*, goes further. He says that the profession of psychology suffers from "physics envy." It characterizes mental disorders as disease using the *Diagnostic and Statistical Manual of Mental Disorders* (DSM) "without providing an actual scientific basis."

What research *has* shown is that serotonin and other neurotransmitters such as dopamine and GABA are linked to mood. The relationship isn't straightforward, with more being better and less being worse, or vice versa. Mood and a sense of well-being has more to do with the *ratios* of neurotransmitters, which is why I use the analogy of a cocktail of chemicals.

Bias against women has also been hindering science for decades. The pandemic of 2020 was one in a string of epidemics involving strains of coronavirus. There are hundreds of them, seven of which are known to infect humans. In 2002, the SARS coronavirus made headlines. In 2012, MERS.

In 1964, the first coronavirus was identified by June Almeida, a virologist working at St Thomas' Hospital in London. But even though she was an electron microscope superstar, the medical journal to which she submitted her paper rejected it "because the referees said the images she produced were just bad pictures of influenza virus particles." Like Betty Twarog, the seminal nature of June Almeida's work was eventually recognized.

While neurotransmitters and hormones are distinct in their derivation, some molecules are able to operate as both. Norepinephrine (aka noradrenaline) is one example. The adrenal glands can release it into the bloodstream as a hormone and the nervous system can release it into synapses as a neurotransmitter.

HORMONE RATIOS, AGING, AND VITALITY

Many hormones are household words. Everyone has some idea of what testosterone is, what it does, and where it's secreted in men's bodies. Women make testosterone too, but in the adrenal glands and ovaries.

Among the other hormones released by the adrenal glands are cortisol, epinephrine, DHEA, and norepinephrine. DHEA is the most common hormone in your body, and it is associated with cell regeneration and communication. Its full name is the tongue-twister dehydroepiandrosterone. You need a PhD to pronounce that, so we'll stick to calling it DHEA.

In my keynote speeches, I often call DHEA the "youthening hormone" because it's associated with many beneficial anti aging effects, including increased muscle mass, improved bone density, and reduced wrinkling of the skin. DHEA deficiency is linked to cancer, diabetes, heart disease, obesity, and other ills.

The molecular structure of DHEA is very similar to that of cortisol, though they produce opposite responses in the body. As with neurotransmitters, it's the *ratio* that is often more important than the *absolute amounts* your body secretes. And you change the ratio simply by stress-reduction practices like tapping and meditation. You're in charge of your own personal anti aging program, simply by the way you deploy your consciousness.

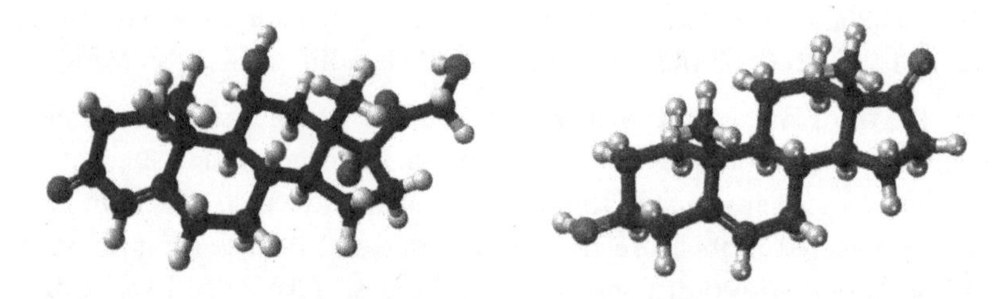

5.5. The structure of the cortisol and DHEA molecules is very similar.

Opium and Stray Dogs

Friedrich Wilhelm Adam Sertürner was born in 1783, the fourth of six children. After the death of his parents, he became a pharmacist's assistant in the German town of Paderborn.

In the early 1800s, he became intrigued by the medicinal properties of opium, a painkiller and soporific used widely by the doctors of his day.

COLOR SUPPLEMENT

CHAPTER 1

1.2. An unedited color photograph taken the day after the fire.

1.3. The same perspective two years before the fire.

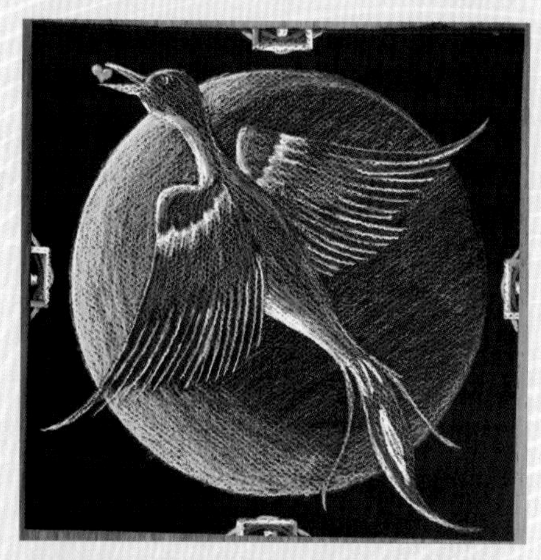

1.15. Healing mandalas.

1.18. Measuring EEG in workshop participants.

Both brain function and structure are changed by sustained meditation. The Default Mode Network (DMN) is composed of two regions in the centerline of the brain. The DMN is self-focused and distracts us from meditation. When active, it recruits many other brain regions, including those that govern cognition and emotion.

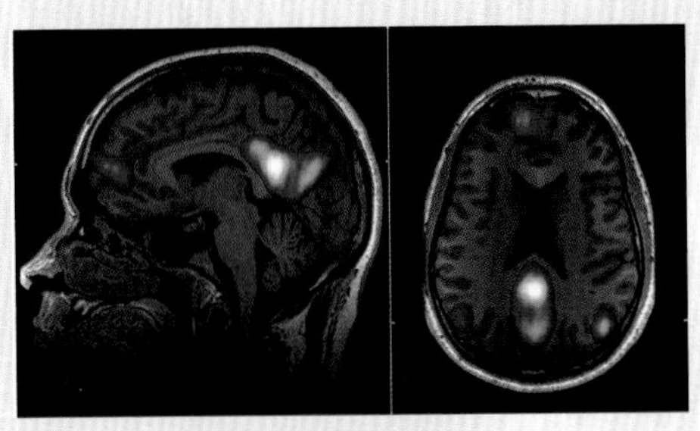

2.4. Brain regions active in the Default Mode Network.

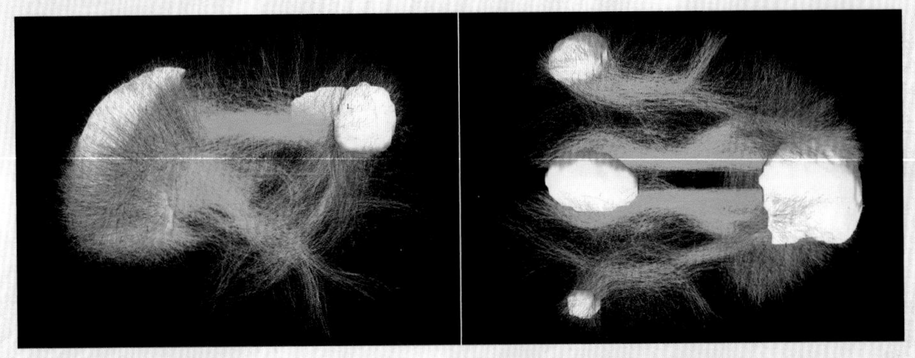

2.5. Nerves from the Default Mode Network reach out to communicate with many other parts of the brain.

2.6. The demon attempts to shake the single-minded focus of the prince.

Evolution gave us three brains stacked on top of each other. The most ancient part is the brainstem and cerebellum, which control the survival functions like respiration, digestion, reproduction, and sleep. It began evolving some 600 million years ago.

Stacked above the hindbrain is the midbrain or limbic system. It handles memory and learning, as well as emotion. It evolved more recently, about 150 million years ago, with the rise of mammals.

The biggest part of the human brain is the neocortex. It is also the newest, in terms of evolutionary development. The brain has tripled in size since early humans began making tools about 3 million years ago. The neocortex handles conscious thought, language, sensory perception, and generating instructions for the other brain regions.

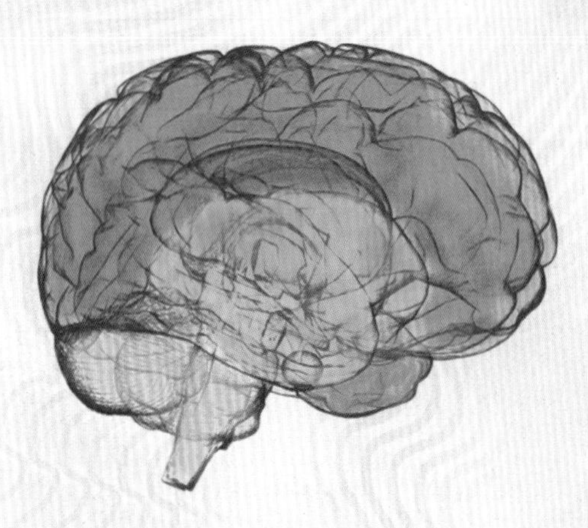

3.2. The three-part brain.

The cortex has four lobes. the frontal lobe, parietal lobe, occipital lobe, and temporal lobe. Each one has distinct functions. The occipital lobe processes information from our eyes. The temporal lobe integrates memory with input from our senses of sound, touch, taste, smell, and sight. Information about movement, temperature, taste, and touch is processed by the parietal lobe. Executive functions like cognition and the control of our actions are the province of the frontal lobe, which is also the seat of what we think of as "self."

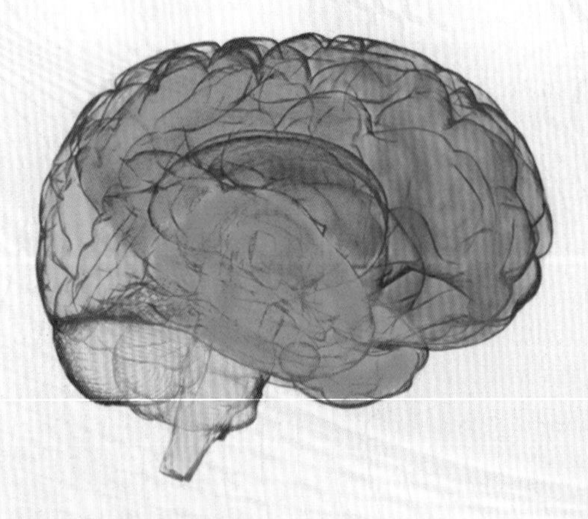

3.3. The four lobes of the cortex.

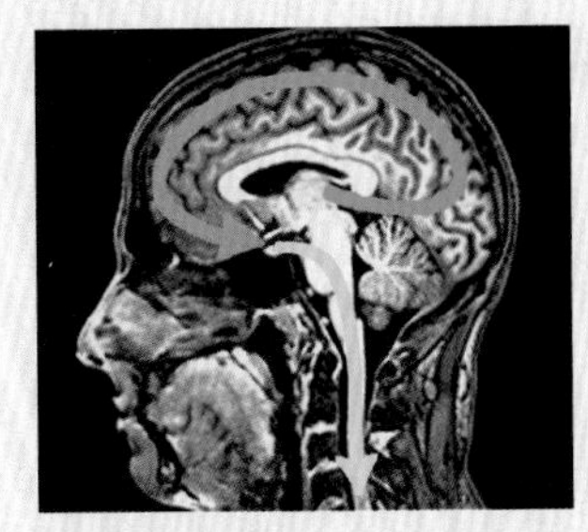

3.8. The long path.

When we receive sensory input, it travels up the spinal cord to the emotional centers of the brain. They query the executive centers like the prefrontal cortex to determine whether or not a stimulus is threatening. If it is not, no signal is passed to the fight-flight system. This loop is called "the long path."

Traumatized people and those with poor emotional regulation often go directly into fight-flight. The emotional centers send the alarm signal to the body directly, bypassing the checks and balances of the brain's cognitive centers. This is "the short path."

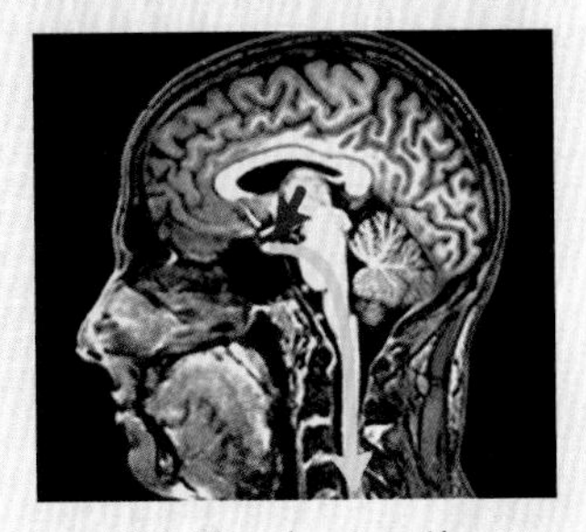

3.9. The short path.

CHAPTER 4

4.1. "Three Nuns in the Portal of a Church" by Armand Gautier.

4.3. "The Eleusinian Mysteries" by Genrich Ippolitovich Semiradsky.

Researchers began using EEGs to record the brain-wave patterns of meditators and healers in the 1960s. They found a unique and distinct ratio of brain waves. It's characterized by large amplitudes of alpha waves, the signature waves of relaxation, and small amplitudes of beta, the wave of stress.

Later, this same "Awakened Mind" pattern was found in peak performers of all types, whether artists, scientists, athletes, or businesspeople. The Awakened Mind is measured by various EEG systems, especially the Mind Mirror, which was developed specifically to identify advanced states of consciousness like Bliss Brain. It is usually hooked up to the brain's occipital and temporal lobes.

4.7. The Awakened Mind pattern.

4.25. Tibetan yantra.

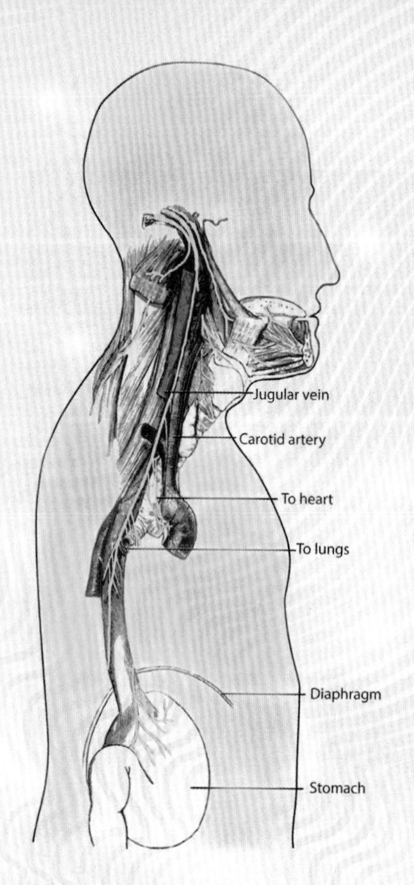

Activation of the vagus nerve produces relaxation. It connects with all the body's major organ systems, so when it is signaling, we feel completely relaxed.

4.8. The vagus nerve connects with all the major organ systems of your body.

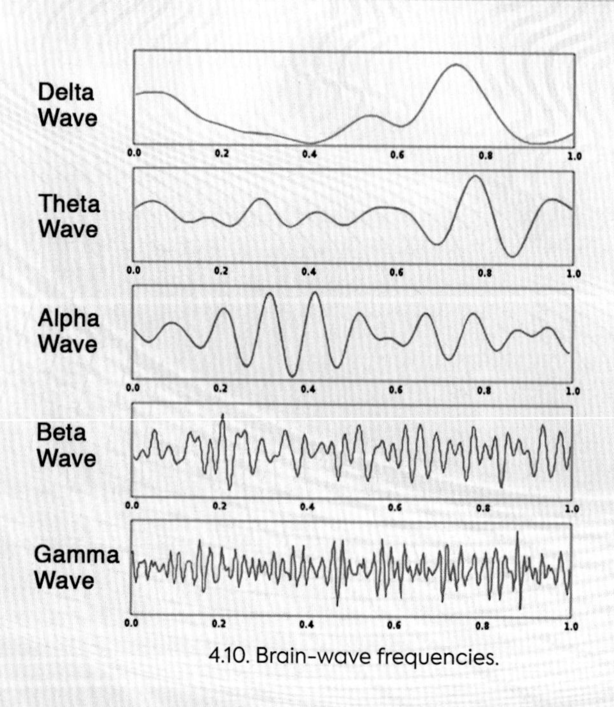

4.10. Brain-wave frequencies.

EEGs measure both the frequency of brain waves (e.g. delta, theta, alpha, beta, gamma) and their amplitude (strength).

Frequency is one of the two ways we measure brain waves. Frequency is the type of brain wave, such as alpha, beta, or delta. The second way, amplitude, is how strong those brain waves are.

When a researcher says that a subject has "bigger" or "increased" or

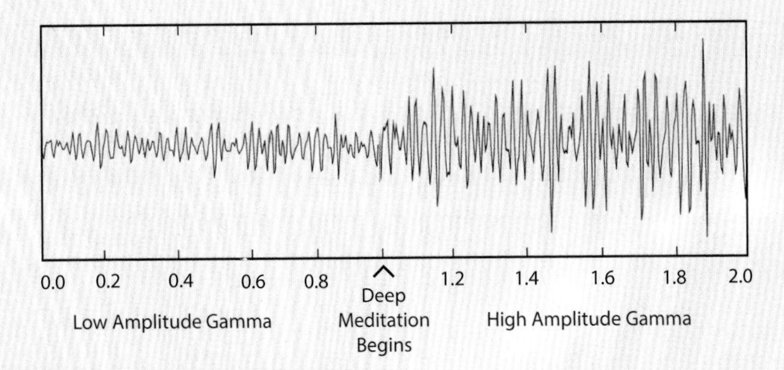

4.11. Brain-wave amplitudes.

"greater" or "more" of a brain wave, we mean greater amplitude. When we say that a brain wave "shrinks" or "reduces" or "lessens," we mean reduced amplitude. The image above depicts a 2 second readout of low and high amplitude gamma.

The Muse is a popular personal EEG device. Unlike the Mind Mirror, it primarily measures activity in the frontal lobe.

4.12. Muse readout of someone in a calm state of ordinary consciousness.

Here's what a typical Muse research readout looks like. From left to right, you see brain-wave frequencies. On the far left is delta and on the far right is gamma, with all the others in between. The dotted white lines running left to right represent 10 seconds of recording.

There are five shades of color on the readout, and they depict brain activity from very high to very low, using the following color scale:

Red = very high activity
Yellow = increased activity
Green = normal activity
Light blue = decreased activity
Dark blue = very low activity

The person whose readout is showing in 4.12 is in an ordinary state of consciousness, just like you and me when we're awake, relaxed, and calm. Brain function is normal, so most of the frequencies are in green. Not too much brain activity, not too little. The time stamp shows that, in each 10-second period, nothing much changes.

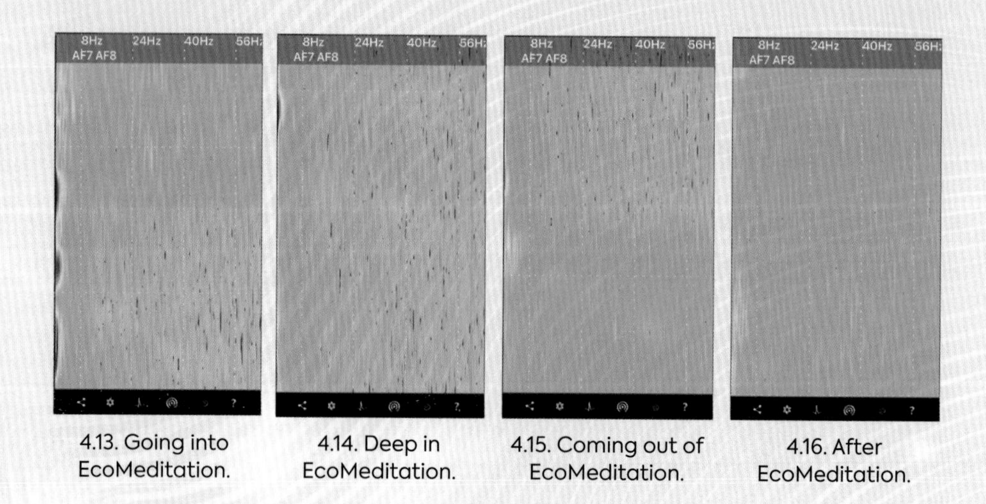

4.13. Going into EcoMeditation.

4.14. Deep in EcoMeditation.

4.15. Coming out of EcoMeditation.

4.16. After EcoMeditation.

In 4.13, you see the changes that occur in the brain of a person beginning an EcoMeditation session. Her brain function changes radically. Within less than five seconds, she's producing more delta and theta waves, the signature waves of intuition, healing, and connection with the universe. That's the oval-shaped flare on the left, in the frequencies of 0 to 8 Hz.

You see a steep drop in all the high brain-wave frequencies, as self-preoccupation turns off. That's the prefrontal cortex (PFC) shutting down. The 40% drop in PFC function measured in meditators kicks in and she's suddenly in a radically altered state of consciousness.

As the meditation experience ends, the brain prepares to return to ordinary consciousness. When the meditator opens her eyes, her brain returns quickly to its normal state.

Yet it's not the same normal that she was in when she started meditation. Andrew Newberg observes that, "The brain that comes back from the enlightenment experience is not the brain that entered it."

Illustration 4.17 shows you before and after, side by side. The first is the "normal" she experienced before she started her EcoMeditation practice. The second is her brain's normal after 30 minutes of meditation. You'll see it's noticeably calmer and more integrated. There are little blips of dark blue even in the midst of the green.

An MRI study at Bond University compared brain function in 25 participants. Half were randomized into a group that listened to an EcoMeditation audio track. A control group did a mindful breathing exercise while recalling a recent vacation.

After listening to their tracks every day for 2 weeks, participants came back to the lab for follow-up tests. There was little change in brain function in the control group. But in the EcoMeditation group, we found significant differences

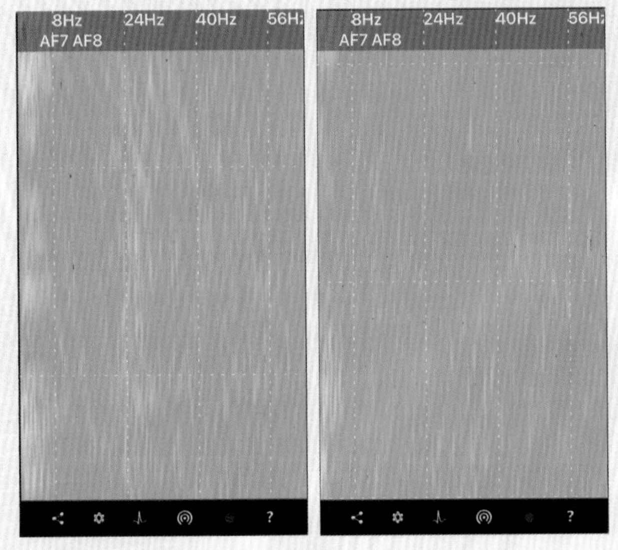

4.17. Before and after, side by side.

in two important brain networks.

The first was an increase in connectivity between the hippocampus and the insula. The hippocampus, seat of emotion, learning, and memory, showed increased connectivity with the right insula, central to kindness and compassion.

Secondly, we found reduced activity in the medial prefrontal cortex, the "seat of the self" and also one of the two regions making up the Default Mode Network. Much of the left prefrontal cortex went offline as well, as the "thinking brain" deactivated.

These findings suggest that compassion increased in those practicing EcoMeditation. This includes both compassion for self as well as compassion for others, since EcoMeditation has you send kind intentions both ways. At the same time, participants benefited from a shutdown of one of the two poles of the DMN, freeing them up from its constant self-absorbed chatter. Mood research shows that when the DMN shuts down, you're much happier.

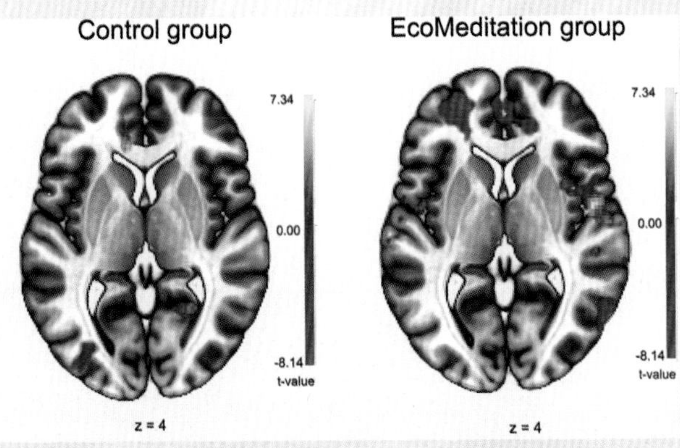

4.18. Changes in connectivity in a group doing EcoMeditation versus a placebo. Blue through purple indicates lessened activity while red through yellow shows heightened activity. The blue area in the front of the brain at the right is the medial prefrontal cortex, one of the two poles of the DMN. The bright red and yellow spots on the right are the insula.

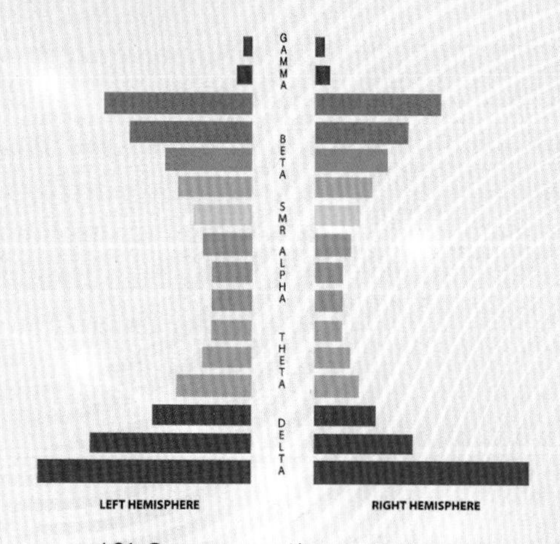

4.21. Gamma are the waves in red at the very top of the screen in this Mind Mirror EEG readout.

When the Mind Mirror EEG is used to measure brain waves, a much richer picture of brain activity emerges than with simple home devices like the Muse. It measures gamma, the fastest wave produced by the brain.

Gamma is the signature wave of the "flow" state and represents the synchronization of information from many different brain regions. It's usually found in highly creative people, as well as ordinary people having a moment of insight.

It's also observed in states of mystical union. When studying advanced yogis, Davidson found that their brains had 25 times the gamma activity of ordinary people.

Gamma is also associated with feelings of love and compassion, increased perceptual organization, associative learning, synaptic efficiency, healing, attention, and states of transcendent bliss. Neuroimaging studies have shown that gamma waves synchronize the four lobes of the brain across frequencies and engender whole-brain coherence.

That's a long list of benefits, all hallmarks of the high-performance brain. And while the studies of Tibetan monks show that big gamma appears only in adepts, EcoMeditation research shows it appearing even in novices.

CHAPTER 5

5.18. The only way to get all the most pleasurable neurochemicals surging through your brain at one time is the ecstatic flow state found in deep meditation.

5.19. Ecstatic mystical experiences are common throughout history. In 1601 Renaissance painter Guiseppe Baglione depicted "The Ecstasy of St. Francis."

CHAPTER 6

The dentate gyrus is a structure in the midbrain that coordinates the regulation of emotion among many brain regions and helps control the DMN.

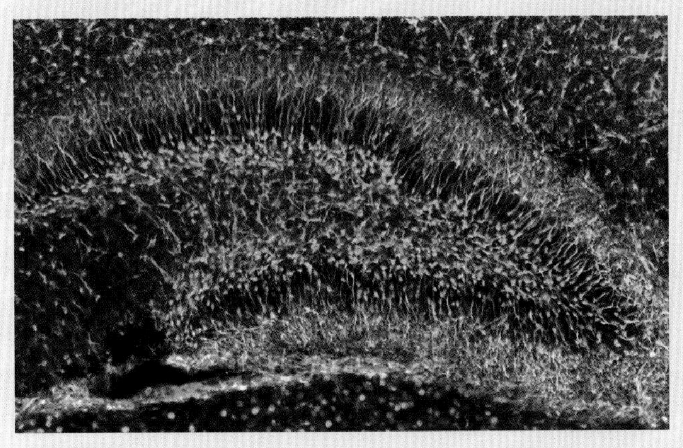

6.8. Neural stem cells in the C-shaped dentate gyrus.

7.3. Methyl groups (bright spheres) adhering to a DNA strand.

Molecules called methyls, which attach to genes, inhibit their expression. Certain molecular messengers are associated with depression, anxiety, and PTSD. This is just one way in which childhood trauma results in "biological embedding" of negative experiences.

7.7. The fire mirror—an art piece created by Christine Church out of fragments of pottery and other objects found in the rubble after the fire.

THE FIRE MIRROR

by Christine Church

The mosaic mirror called Fire Fragments tells a story of my past. This is an assemblage I made from pieces of pottery and jewelry. As I found each item in the burned rubble, I would wonder, "Out of thousands of things, why did this piece make it?"

Or "Look where we found this!" Objects that once had little or no meaning now became fellow survivors: symbols of strength and courage.

Top right: We never found the diamonds I was looking for, but we found a brown mass of jewelry melded together. In it was my mom's wedding ring! We threw piece after unrecognizable piece into a bucket. Our friend, Ray even brought his gold-panning equipment. We were so crazed that we didn't even use our masks or gloves.

Moving counterclockwise: My great-grandmother's Bavarian china, two Spanish Lladros from my dad's collection—"Girl with a Shoe" and "The Shoe" (top left and bottom left corners) with Bavarian china cup and saucer.

The cat inside the shoe represents our beloved Apple and Pierre. There is a Moroccan necklace that I wore in my 20s when I traveled to Spain and Morocco (bottom) and a duck made from the ashes of the Mt. St. Helens volcano.

Hope arises in the bottom right corner with a cup from a ceramic tea set my daughter Julia made in seventh grade, with a piece of melted glass inside. Glass melts at 2,500 degrees—that's how incredibly hot the fire burned!

The key to our house (right side) is in a symbolic pond where our nameless turtle found her name when we discovered she had survived the fire. Tubbs now lives out her days in a friend's backyard pond.

Rings and shards of pottery I had made are scattered throughout. A seed and a crystal from our new home are planted (lower right corner). Angels watching over us with love that night and always (top right and left corners).

I feel it was a blessing that these broken, burned pieces from the ashes could be repurposed into an art form. The process of putting it all together was a healing, and I felt a sense of completion and letting go when it was finished. It hangs in our house as a reminder of things lost but not forgotten. And turning disaster into beauty.

CHAPTER 8

8.2. Identical twins look the same at birth.

8.3. But they diverge over the years, and may look quite different at age 60. While chronologically the same age, biologically some twin pairs are more than 10 years apart.

8.5. New life forms that emerged during the Cambrian explosion.

Climate science tells us that we're facing a potential 3-foot (1-meter) rise in sea levels this century. In one low-lying country, Bangladesh, a 3-foot rise will put 20% of the country underwater, and 30 million people will become climate refugees.

The average elevation of the land in the US state of Florida is 6 feet (2 meters) above sea level. Sea level rises of 10 feet (3 meters) are predicted between 2050 and 2075, and 50 feet (15 meters) by 2500.

8.11. Map of the Earth with a 6 meter (20 ft) sea level rise represented in red.

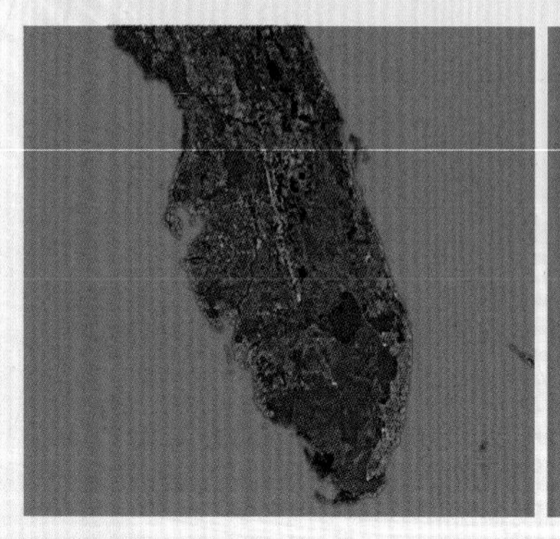

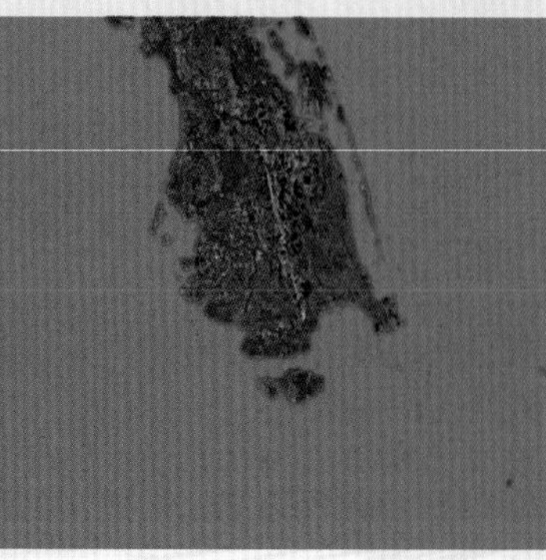

8.13. Map of the US state of Florida showing areas above water after the 3 meter (10 ft) rise predicted 2050 - 2075.

Sertürner's curious mind wanted to determine which substance in the opium poppy was the active ingredient. Experimenting on stray dogs and his own body, he eventually isolated an alkaloid molecule from the poppy's resin. He named the compound "morphine," after Morpheus, the Greek god of dreams. He published his findings in 1803, with a second and more comprehensive description in 1817.

With the development of the hypodermic needle in 1853, morphine had an efficient vehicle for delivering the drug into the human circulatory system.

Doctors quickly became aware of its addictive properties, however, and in 1914, the US Congress passed the Harrison Narcotics Act, the precursor of 1971's Controlled Substances Act, to regulate the manufacture and distribution of addictive substances.

EXOGENOUS AND ENDOGENOUS CHEMICALS

This swirling cocktail of chemicals is being synthesized inside our bodies moment by moment. Endorphins, for instance, are a family of neurotransmitters that block pain. When we make more endorphins, we block the sensation of pain.

Many drugs are specifically designed to *mimic the action of our body's natural chemicals*. Morphine is an example. Developed to dull the sensation of pain, it mimics the action of our brain's natural endorphin molecules.

Endorphins are "endogenous," meaning that they are produced inside the body. Morphine is "exogenous." That's just a 10-dollar word for "external." We can get neurochemicals such as painkillers from inside our own brains and bodies or externally by using pharmaceutical drugs.

Exogenous molecules like opioids, alcohol, morphine, and prescription drugs are effective because they fit into the same receptor sites as the endogenous molecules made by our own bodies. Each is like a key that fits into the same lock. It may not be identical to the endogenous molecule, but it is close enough to dock to a receptor and unlock the same physical response.

THC (tetrahydrocannabinol), the active ingredient in marijuana, functions in this way as well. We'll see in this chapter how THC mimics the action of anandamide, our brain's endogenous "bliss molecule."

Not only can our brains synthesize our own personal THC, we can use our consciousness to generate the mood-altering molecules produced by cocaine, MDMA (aka "ecstasy"), morphine, ayahuasca, psilocybin (magic mushrooms), and alcohol.

All at once! The amazing research in this chapter will show you how to get seriously happy.

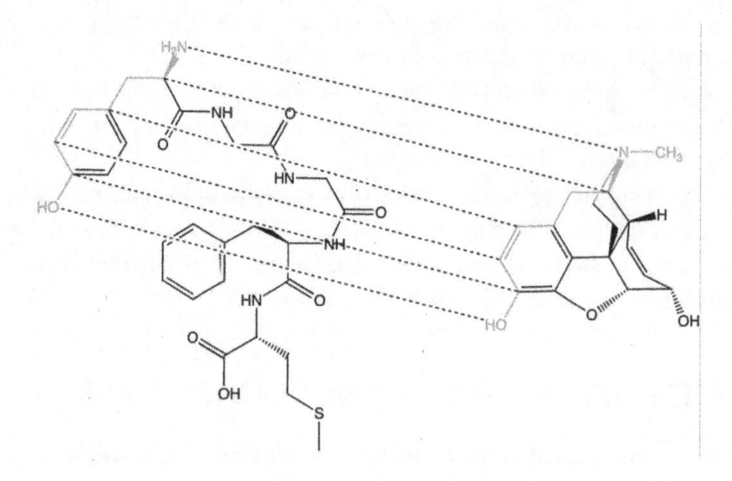

5.6. The reason that morphine (right) is effective is that it has a molecular structure similar to that of an endogenous opioid molecule (left), allowing it to fit into the same receptor lock.

The most popular legal drug in the world is alcohol. It triggers the release of the inhibitory neurotransmitter GABA. At the same time, it suppresses the action of the excitatory neurotransmitter glutamate. This combination works quickly to produce a feeling of euphoria, the "buzz" associated with drinking.

Other drugs that stimulate GABA include muscle relaxants and benzodiazepines. One of the best-selling examples is carisoprodol, sold under the enchanting brand name of Soma. Xanax and Valium are among the best-selling benzodiazepines.

Alcohol also increases serotonin and, when drunk in large quantities, the most powerful of painkillers, beta-endorphin. After a few shots, the drunk declares truthfully, "I feel no pain." Heroin and prescription opioids also dock with the same receptors as endogenous endorphins like beta-endorphin. Cocaine engages the dopamine neurotransmitter system, while nicotine acts through the neurotransmitter acetylcholine.

So the reason that all these exogenous molecules can make us feel good is that they mimic the action of endogenous feel-good molecules.

Two Paths to Seventh Heaven

"Why are you here?" asked Dr. Costello, the brown eyes below her surgical cap sparkling with intelligence.

"To give and receive love! To connect with the universe with a huge heart and open mind in every moment!" I replied enthusiastically.

She looked startled. It was a few weeks after the Tubbs fire, and I was in pre-op to repair an injury I had sustained in the aftermath. I'd lucked out with a superb surgeon, Dr. Stewart, who had just patiently reviewed the details of the procedure with me again.

Now it was the turn of the anesthesiologist, Dr. Costello. I guessed that her question was intended to make sure that the hospital was matching each patient with the correct procedure. If you're in the hospital for cataract surgery, it would be a crying shame if you received a knee replacement by mistake. Hence the question about why you're here.

As she recovered from her surprise and opened her mouth to rebuke my levity, I forestalled her with, "I'm here for a right inguinal hernia repair. Can you tell me about the anesthetic I'll receive?"

"It's going to make you feel very, very good," she replied, as she busied herself connecting the drip.

And indeed it did. In a sea of bliss, I passed out a few moments later. Five hours later, I was home after a successful operation.

What I had not expected was that I would feel so good afterward. That evening, I was still floating on the sea of bliss, all the way into bedtime. I woke up the next morning still feeling ecstatic. Christine was alarmed. It was great that I was feeling no pain. But to be in Seventh Heaven? That seemed unnatural.

The effect wore off on the third day and the pain kicked in. Back down to earth.

I now understood why people get addicted to opioids. That feeling of ecstasy could be produced by simply popping a pill. For the first time, I'd experienced Bliss Brain from an exogenous source. It was sure a lot less toil than putting in 10,000 hours of meditation.

Meditation made a difference after the operation. I began the third day's meditation in pain, but as endorphins flooded my brain, the pain disappeared. I tuned in to Bliss Brain several times a day and I never had to open the bottle of painkillers I'd been prescribed. I tapped and affirmed a lightning-fast recovery.

Dr. Stewart was indeed surprised when he examined the surgical site a week later. Pain-free and super-fast recovery. Brain drugs rule!

THE SEVEN DRUGS OF BLISS

Chapter 4 showed that when we meditate, our brain waves, indicators of neural activity, change dramatically. Stress-induced beta waves drop in amplitude. Healing theta waves amplify, as do connection-inducing delta waves. Our highest-frequency wave, gamma, "rides" on big delta, so we then start to produce the "genius wave" of gamma in large amounts. Meditation trains the brain to create an "alpha bridge" between the conscious and superconscious frequencies of beta and gamma, and the subconscious and unconscious frequencies of theta and delta.

Just as the stress-filled wave of beta is associated with a rise in the stress hormone cortisol, these other waves are also associated with hormone and neurotransmitter changes. Bliss Brain triggers the release of the chemicals of ecstasy. Science has so far identified at least seven of these.

Here we'll review them one by one. Individually, they make us feel good. But the *combined* cocktail produced by meditative flow states can transport the brain into bliss.

Dopamine: The Motivational Neurotransmitter

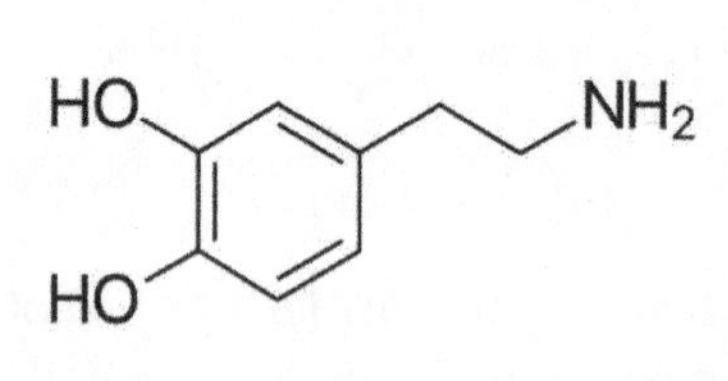

5.7. Dopamine: your motivational molecule.

The release of dopamine in the brain is part of the "reward system." It's like a kid in school getting a lollipop for giving the right answer on a quiz.

Right answer = lollipop. Wrong answer = no lollipop. Dopamine is linked to theta brain waves.

Two areas of the brain are integral to the dopamine reward system: the tegmentum near the brain stem and the nucleus accumbens. The latter, which we encountered in Chapter 3, is associated with reward and motivation, and some scientists consider it part of the emotion- and learning-focused limbic system.

Pleasure triggers tegmental neurons to release dopamine. That pleasure can take many different forms, both external and internal, both emotional and physical. Big-time pleasures such as sex stimulate the release of dopamine. Even thinking about sex turns on the dopamine reward system. Likewise, *anticipation of pleasure* of other kinds, such as financial reward, can activate the dopamine pathway.

Mind-altering substances such as alcohol, heroin, and cocaine stimulate dopamine release in the nucleus accumbens. When an addict even *thinks about* getting high, dopamine kicks in. On the opposite end of the spectrum from pleasure and reward, dopamine levels can be depleted by chronic pain or chronic stress.

Robert M. Sapolsky, Stanford professor of biology and neurology and author of the brilliant tome *Behave: The Biology of Humans at Our Best and Worst,* describes the special nature of dopamine, "Dopamine is not just about reward anticipation. It fuels the goal-directed behavior needed to gain that reward . . . In other words, dopamine is not about the happiness of reward. It's about the happiness of pursuit of reward."

Serotonin: The Feel-Good Neurotransmitter

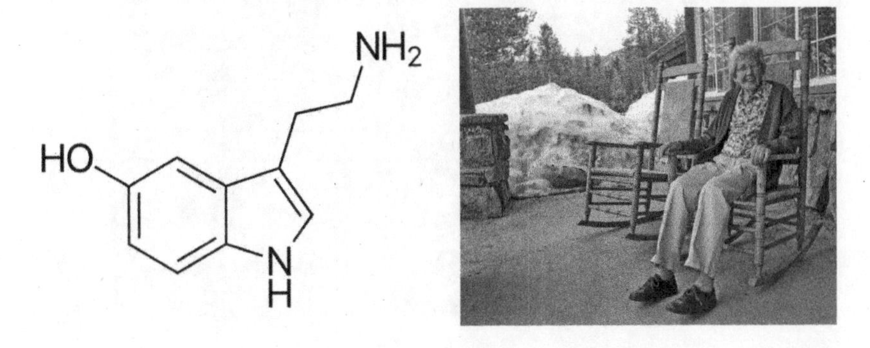

5.8. Serotonin: your satisfaction molecule.

Most serotonin is produced in an area of the brain stem called the raphe nucleus. Serotonin is then sent to the nucleus accumbens, the tegmentum, the amygdala, and the prefrontal cortex, working with dopamine there to reinforce goal-oriented behavior. Motivate yourself with dopamine, achieve a goal, and you get a rush of the feel-good chemical serotonin.

High serotonin levels are associated with feelings of both mental and physical well-being. On the flip side, low serotonin is linked to greater aggression and impulsivity in both thought and behavior, up to and including the ultimate feel-bad act of suicide. Serotonin is found in association with alpha brain waves, and it is believed to mediate the ratio between alpha and the two slowest frequencies of theta and delta. Here again we see the importance of *ratios* rather than *absolute amounts.*

People had been seeking the feel-good effects of serotonin long before Betty Twarog's discovery in 1952. The chemical structure of serotonin is similar to that of psilocybin, a psychedelic compound that induces euphoria and hallucinations. It is found in over 200 species of mushroom, hence the nickname "magic mushrooms."

Tripping through Time on Magic Mushrooms

For millennia, human beings have been using mushrooms to get high, mimicking the effects of the neurotransmitter serotonin.

5.9. Mayan mushroom sculptures.

The earliest evidence dates back 9,000 years. Rock paintings in North Africa from that period depict mushrooms, a form of "sacred medicine" from that time. In nearby Spain, rock art from Selva Pascuala also shows a row of mushrooms.

Far away in South America, psychedelic mushrooms were likely used by the Maya and Aztecs. The Aztec word *teonanácat*, translated as "flesh of the gods," is believed to have been a hallucinogenic mushroom used in sacred ceremonies. An 1,800-year-old statue discovered in Mexico resembles a mushroom in the *Psilocybe* family.

Mushrooms appear in the art of many other cultures, including the ancient Greeks. A marble grave decoration from around 460 BC depicts two women, heads bent together in a posture of intimate rapport, holding plants that resemble mushrooms.

5.10. Ancient Greek marble grave decoration from Pharsalos, Thessaly.

In the late 1950s, the Western world discovered psilocybin. Gordon Wasson was a banker and part-time mycologist. On a trip to Mexico, he and his wife, Valentina, met Maria Sabina, a Mazatec *curandero*, or herbalist. She shared the secrets and the experience of the sacred mushroom with the couple. They wrote up their experiences for *Life* magazine.

Their 1957 article "Seeking the Magic Mushroom" caused a sensation. Wasson wrote: "The mushroom permits you to see, more clearly than our perishing mortal eye can see, vistas beyond the horizons of this life, to travel backwards and forwards in time, to enter other planes of existence, even (as the Indians say) to know God."

Flood the brain with a serotonin substitute and the tide can wash you all the way up to heaven. No wonder human beings from Stone Age Africans to modern trippers have treasured this potent stimulant.

Norepinephrine: The Wake-Up Neurotransmitter

One of norepinephrine's effects on the brain is to *sharpen attention*. As we saw earlier, norepinephrine (aka noradrenaline) can function as both a neurotransmitter and a hormone. When we perceive stress and activate the fight-or-flight response, the brain produces bursts of norepinephrine, triggering anxiety.

But sustained and moderate secretion can also produce a beneficial result in the form of heightened attention, even euphoria, and meditation has been shown to produce a rise in norepinephrine in the brain. A modest dose of norepinephrine is also associated with reduced beta brain waves.

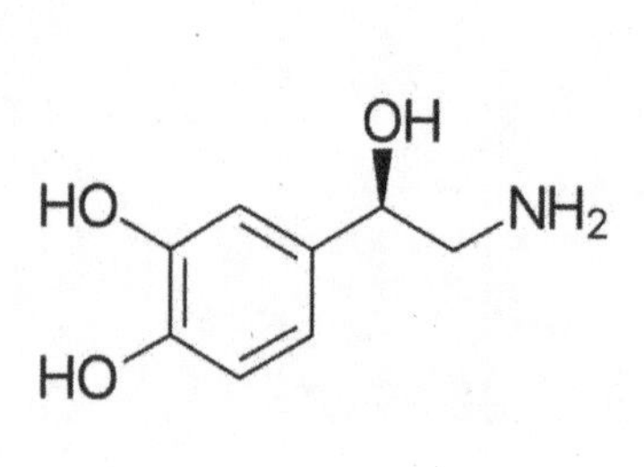

5.11. Norepinephrine: your wake-up molecule.

Notice the paradox here. Norepinephrine is associated with both anxiety and attentiveness. How do you get enough to be alert, but not so much you're stressed?

Surrender is the key.

Steven Kotler, co-author of *Stealing Fire*, says that stress neurochemicals like norepinephrine actually prime the brain for flow states. At first, the meditator is frustrated by Monkey Mind. But if she surrenders, despite the perpetual self-chatter of the DMN, she enters the next phase of flow, which is focus. She has hacked her biology, using the negative experience of mind wandering as a springboard to flow.

Norepinephrine's molecular structure is similar to its cousin, epinephrine. While epinephrine works on a number of sites in the body, norepinephrine works exclusively on the arteries. When both dopamine and norepinephrine are present in the brain at the same time, they *amplify focus*. Attention becomes sharp, while perception is enhanced.

Staying alert is a key function of the brain's attention circuit, which keeps you focused on the object of your meditation and counteracts the wandering mind. It also stops you from becoming drowsy, an occupational hazard for meditators. That's because pleasure neurotransmitters such as serotonin and melatonin (for which serotonin is the precursor) can put you to sleep if not balanced by alertness-producing norepinephrine. Again, the ratios are the key.

Oxytocin: The Hug Drug

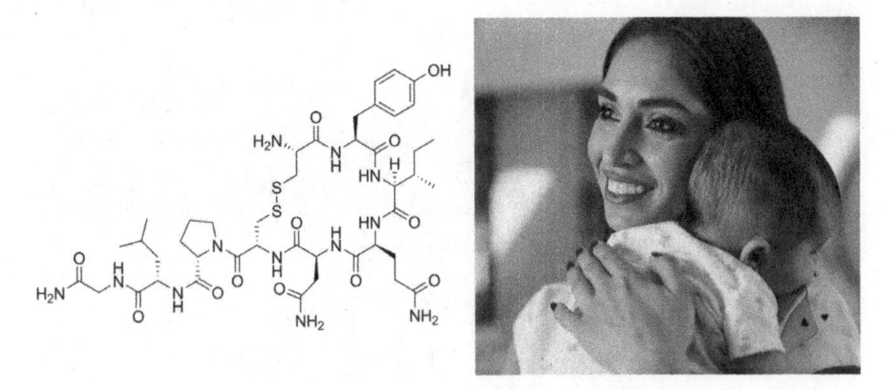

5.12. Oxytocin: your cuddle molecule.

Oxytocin is produced by the hypothalamus, part of the brain's limbic system. When activated, neurons in the hypothalamus stimulate the pituitary gland to release oxytocin into the bloodstream. So even though oxytocin is produced in the brain, it has effects on the body as well, giving it the status of a hormone. It is one of a group of small protein molecules called neuropeptides. A closely related neuropeptide is vasopressin. All mammals produce some variant of these neuropeptides.

Oxytocin promotes bonding between humans. It is responsible for maternal feelings and physically prepares the female body for childbirth and nursing. It is generated through physical touch but also by emotional intimacy. Oxytocin also facilitates generosity and trust within a group.

Oxytocin is the hormone associated with the long slow waves of delta. A researcher hooking subjects up to an EEG found that touch stimulated greater amounts of delta, with certain regions of the skin being more sensitive. The biggest effect was produced by tapping the cheek, as we do in EFT. It produced an 800% spike in delta.

Couples newly in love experience spikes in oxytocin. That's why they touch frequently and have sex often. Oxytocin levels are linked to a man's ability to have an erection. It's a hard drug, so to speak.

5.13. Couples newly in love produce high amounts of oxytocin.

Social groups spike oxytocin. My wife, Christine, paints pictures every Tuesday with her "art tribe," bonding over acrylics, canvas, tea, and oxytocin. A group of teenage girls giggling and gossiping is a veritable oxytocin factory. Whether it's the Hells Angels playing pool in a bar, a Bible study group at a church, or even the Facebook group for Badly Stuffed Animals (11,000-plus users), activities that produce a tribal sense of belonging unlock the synthesis of oxytocin.

The Orgasmic Brain

With functional magnetic resonance imaging (fMRI) and positron emission tomography (PET) technology, we are able to see what goes on in the brain during orgasm. This presents a rich picture of which parts of the brain light up, which shut down, and which hormones accompany each stage of the process. Here is what researchers have discovered:

- Your thinking brain goes offline during sex. Specifically, the lateral orbitofrontal cortex, the region of the brain that regulates decision-making, reasoning, and value judgments, shuts down, taking with it fear and anxiety.
- During orgasm, the areas that light up are the hypothalamus (oxytocin producer), thalamus (integrates touch, movement, and sexual thoughts), and substantia nigra (dopamine producer).
- Orgasm is accompanied by a release of dopamine, oxytocin, and prolactin. Oxytocin triggers feelings of closeness, while prolactin produces the satisfied feeling people experience after orgasm. It also regulates milk production after childbirth.
- Both male and female brains experience the effects of oxytocin during sex, but women continue to release oxytocin after orgasm. This may be why women like cuddling after intercourse.
- Serotonin is released in the brain after orgasm and promotes relaxation. The fact that serotonin is a close cousin of melatonin, the sleep neurotransmitter, may explain why many men fall asleep after sex.
- The way in which your brain is stimulated by orgasm resembles the way it activates when you listen to your favorite music, do drugs, drink alcohol, or gamble. The same reward pathways are activated.
- Orgasm increases blood flow in the brain, which helps keep it healthy.
- Neuroplasticity allows the brain to create new pathways for orgasm independent of the sexual organs. When people are paralyzed from the waist down, for instance, they can be retrained to experience orgasm via the stimulation of another part of the body, such as the skin of the arm.

Touch doesn't just feel good. It can produce invisible but pervasive healing effects in your body. Oxytocin deactivates glutamate, the most excitatory of neurotransmitters, making you feel calmer and more emotionally balanced. It stimulates the release of nitric oxide in your blood vessels, improving circulation.

To test the physical effects of oxytocin, a group of researchers administered psychological and physical tests to 34 married couples. Half were randomized into a control group that kept a diary of their moods and levels of physical affection.

The experimental group were taught a technique called "listening touch." They tuned in to their partner's mood by touching the partner's neck, shoulders, and hands. They did this both back-to-back and facing each other. Men and women alternated giving and receiving "listening touch." They practiced for 4 weeks at home, 30 minutes three times per week, then went back into the lab for a second set of tests.

The researchers found that touch was producing oxytocin as expected. But as the hormone flowed through the blood vessels of the husbands, it triggered the production of nitric oxide. This lowered their blood pressure and the beneficial effects persisted for a whole day.

Oxytocin can even produce bonding between species. One study demonstrated that when a person and a dog interact, oxytocin is released in both species. The levels increase as they gaze lovingly at each other, just like a human couple in love. The longer they gaze, the more oxytocin is released, stimulating yet more gazing.

Beta-endorphin: Nature's Own Pain Reliever

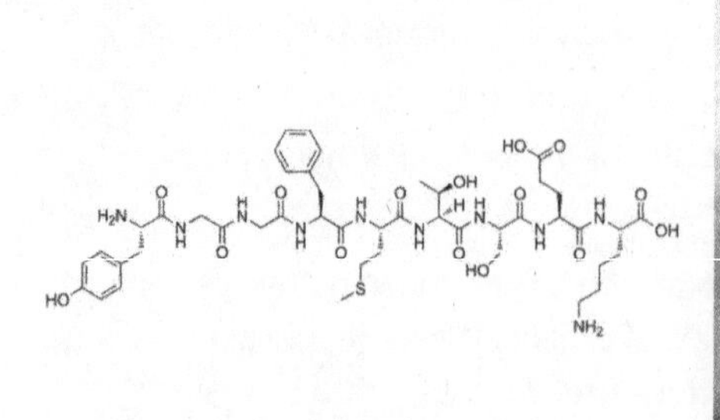

5.14. Beta-endorphin: your painkilling molecule.

The pituitary gland produces the powerful pain reliever known as beta-endorphin, which is also a neuropeptide. Endorphins dock with opiate receptors in the brain and diminish or block our perception of pain. The pharmaceutical drugs morphine and codeine work the same way. Endorphins are accompanied by alpha and theta brain waves.

One study found that when injected directly into the brain, beta-endorphin is over 17 times more powerful a painkiller than synthetic morphine. Another study found that when injected into a vein, beta-endorphin is three times stronger. That's a very powerful pain reliever, and the brilliant pharmaceutical company that manufactures it is Your Neurons, Inc.

Anandamide: The Bliss Molecule

5.15. Anandamide: your bliss molecule.

Anandamide is known as the joy chemical. It has also been called "the brain's own marijuana" and "the bliss molecule." Research has demonstrated that anandamide is involved in the neural generation of pleasure and motivation. We'll talk about this a lot in the section called "Getting High: The Endocannabinoid System and Anandamide."

Anandamide is associated with high-amplitude theta waves and the bursts of gamma we identified in Chapter 4. Gamma waves are sustained for long periods of time in seasoned meditators, as we'll discover in Chapter 6.

Nitric Oxide: The Oxygen Bomb

$N=O$

5.16. Nitric oxide: your energy molecule.

Nitric oxide is both a free radical and a gas, manufactured in the body from plant nitrates in the diet combined with antioxidants such as vitamin C. It's not to be confused with nitrous oxide, aka laughing gas.

The importance of nitric oxide in the context of our discussion on substances that affect the brain lies in its ability to improve brain neuroplasticity. It does this by enhancing oxygen flow to the somatomotor cortex, a region of the brain vital to learning and memory.

While most free radicals are detrimental to health, nitric oxide in normal amounts is not. Among the many benefits of nitric oxide are promotion of normal blood pressure, protection against heart attacks, boosting of immune function, antibacterial action, suppression of inflammation, maintenance of fluid balance, and protection against diabetes. Nitric oxide is associated with alpha brain waves.

THE RELATIONSHIP OF BRAIN WAVES AND NEUROTRANSMITTERS

There's no need to remember the details of these molecules. Simply understand that changes in the energy of the brain are associated with matter in the form of the "molecules of emotion," hormones and neurotransmitters. The process is unimaginably complex, and this isn't a reference book intended to make definitive and comprehensive pronouncements about biochemistry; instead, it is a layperson's guide that uses rule-of-thumb generalities.

That said, the table below presents a rough guide to the associations between the energy flowing in the brain in the form of brain waves and the chemicals that accompany them.

Brain Wave	Neurotransmitter
Gamma	Anandamide
Beta	Cortisol, Norepinephrine
Alpha	Nitric Oxide, Endorphins, Serotonin
Theta	Dopamine
Delta	Oxytocin

5.17. Associations between brain waves and neurochemicals.

GETTING HIGH: THE ENDOCANNABINOID SYSTEM AND ANANDAMIDE

Cannabinoids are psychoactive compounds that provide both medical benefits and the feeling of being high. Plant-derived cannabinoids, such as the THC found in marijuana, are called phytocannabinoids. Few people are aware that *our bodies also produce cannabinoids naturally.* These are called endocannabinoids to distinguish them from their plant-derived relatives.

As with neurotransmitters, our brains have receptors specific to cannabinoids—the lock and key metaphor again. The endocannabinoids and their receptors comprise what is known as the endocannabinoid system. This system plays a role in appetite, pain, inflammation, sleep, stress, mood, memory, motivation, and reward.

There are two types of cannabinoid receptors: type 1 (CB1) and type 2 (CB2). The type 1 receptors are primarily in the brain, and are especially dense in the basal ganglia, cerebellum, and hippocampus. These brain regions are involved in limb control, posture and balance, and learning and memory. CB2 receptors are found in immune tissues such as the tonsils and spleen.

Anandamide is the primary endocannabinoid. It binds to both CB1 and CB2. But while THC is a robust molecule that doesn't break down easily, anandamide does. That's why the natural high wears off quickly while the exogenous one endures.

Low endocannabinoid levels in the human body have been linked to major depression, generalized anxiety disorder, PTSD, multiple sclerosis, attention deficit/hyperactivity disorder (ADHD), Parkinson's disease, fibromyalgia, and sleep disorders.

The health benefits of activating your endocannabinoid system are extensive. They include:

- Stress control
- Anxiety reduction
- Increased optimism
- Improved concentration
- Drops in hyperactivity
- Lower cortisol
- Increased neurogenesis
- Thicker myelin insulators around nerve cells
- Improved sleep
- Reduced impulsivity
- Mood improvement
- Reduced inflammation
- Fewer headaches
- Decreased activity in the amygdala, the brain's fear relay

Among the methods for boosting your endocannabinoid system are eating plenty of fruits and vegetables; the flavonoids they contain slow the breakdown of cannabinoids. Particularly potent endocannabinoid boosters are olive oil (it upregulates CB1 receptors), green tea (it activates cannabinoid receptors), probiotics (they stimulate the endocannabinoid system), dark chocolate (it contains anandamide), and foods rich in omega-3 fatty acids (they upregulate both CB1 and CB2).

The endocannabinoid systems can also be boosted by your behaviors. These include reducing stress (upregulates CB1 receptors), exercising (activates the endocannabinoid system), and deep meditation (raises your level of anandamide).

Besides meditation, one of the funnest ways to boost anandamide is through touch. Both emotional and physical intimacy trigger oxytocin. In a fascinating series of experiments, Don Wei and colleagues at the University of California–Irvine demonstrated that oxytocin spikes anandamide. Engage in social behavior and "oxytocin-driven anandamide signaling" kicks in. What a bargain—two pleasure neurochemicals for the price of one!

Lumír Hanuš and His Bliss Molecule

In the early 1990s, after years of research on the healing qualities of cannabis in his native Czechoslovakia and in the US in association with the National Institutes of Health, analytical chemist Lumír Hanuš, PhD, received an invitation that would change the course of neuroscience.

Professor Raphael Mechoulam, the world's leading expert on cannabinoids, invited Hanuš to join his team at Hebrew University in Israel, where they were conducting groundbreaking research on the cannabis plant. Earlier, Mechoulam had isolated THC (tetrahydrocannabinol), the compound in marijuana that produces the high.

Though Hanuš planned to come to Jerusalem for only a year, he ended up staying permanently. William Devane, an American expert in molecular pharmacology who had discovered cannabinoid receptors in the brain, was part of the research team as well.

Hanuš thought he would be continuing the research he had been doing on the plant's healing properties, but Professor Mechoulam informed him that his target would be to find a ligand or molecule in the brain that binds to cannabinoid receptors.

Hanuš recalls, "When he told me this, I almost fell. I said to myself, 'It's impossible to find the needle in the heap of hay.' But we started and worked intensively one year."

Here Mechoulam had asked a key question. Why should a molecule derived from a plant, cannabis, affect a human being? After all, the evolution of plants and animals diverged millions of years ago.

The logical answer was that there must be a very similar molecule in the human body, along with the ligands to read and interpret it.

At the end of that year, on March 24, 1992, Hanuš and Devane found the needle in the haystack: the first-known endocannabinoid in the human brain. Devane dubbed it anandamide, from the Sanskrit word *ananda*, meaning "joy" or "bliss."

To get the right answers, you have to ask the right questions. Though it seemed at first to Hanuš that looking for a single molecule was like looking for a needle in a haystack, it took only a year after Mechoulam asked him the right question to find it.

The discovery of anandamide was a landmark moment in brain science. It showed that *the brain manufactures molecules that are virtually indistinguishable from psychotropic drugs.* We're getting them endogenously through the *same molecular delivery systems* that administer exogenous drugs.

Cannabinoid receptors were subsequently discovered in parts of the body other than the brain. These discoveries also showed that drugs have their effects because of their similarity to endogenous molecules. They demonstrated that *the elevated emotional states described by mystics weren't just subjective fantasies*; they were grounded in *objective molecular interactions* that could be measured and quantified.

INGREDIENTS OF THE BLISS BRAIN COCKTAIL

Research has shown that *each one of the seven bliss neurochemicals is associated with meditation.* A review and synthesis of the research literature found increases in serotonin, GABA, vasopressin, and melatonin. The dopamine levels of meditators rose by 56%.

Cortisol dropped, and norepinephrine declined to levels appropriate to focused attention without anxiety. The rhythms of the brain's production of beta-endorphins changed. Heightened oxytocin mobilized the synthesis of anandamide in the nucleus accumbens.

A number of studies and reviews show that meditation stimulates the production of nitric oxide, providing meditators with the health benefits of better circulation and brain neuroplasticity. Nitric oxide release is closely coupled with anandamide production; thus meditation and other stress-reducing activities may stimulate the synthesis of both together.

Anandamide can also improve cognitive function, motivation, learning, and memory, while triggering the growth of neurons in the brain centers that govern those functions. A blissed brain is a learning brain; meditation cements our feel-good experiences into brain hardware through increased neuroplasticity. Anandamide also relieves anxiety and depression while stimulating closeness and connection with others.

The scientific literature shows that oxytocin is increased by meditation. As we saw earlier, oxytocin triggers the release of nitric oxide and anandamide, providing the meditator with a trifecta of pleasurable brain chemicals.

5.18. The only way to get all the most pleasurable neurochemicals surging through your brain at one time is the ecstatic flow state found in deep meditation. See full-color version in center of book.

Each of these neurochemicals is pleasurable in its own right, and you can get them from activities that stimulate their production. These activities might get you one or two but *not all seven in one package.*

For instance, the addict craves the hit of dopamine that bathes his basal ganglia when he pops a pill, but the alertness normally provided by norepinephrine is dulled. The hang glider gets an epinephrine high from jumping off the Jungfrau mountain in Switzerland, but the serenity of serotonin eludes her.

The mom gets an oxytocin rush from cuddling her toddler but misses out on the motivational benefits of dopamine. The weight lifter doing HIIT (high-intensity interval training) experiences a pleasurable "nitric oxide dump" through extreme exertion, but there's no burst of insight-producing anandamide in the mix. The triathlete gets a "runner's high" and forgets his sore knees as beta-endorphin kicks in, but he's not getting the bonding associated with oxytocin.

There's only one activity that stimulates the brain to produce all seven at the same time, and that's the ecstatic state of flow. The shortest way there is deep, alpha-driven meditation. When you blend all seven into a single cocktail, the result is euphoria.

Let's see: What might a combination of the first letters of each drug look like? **S**erotonin, **O**xytocin, **N**orepinephrine, **D**opamine, **A**nandamide, **N**itric **o**xide, and **Be**ta-endorphin? Just for fun, let's combine them, and call our cocktail's special blend SONDANoBe. This is the magic formula that, produced inside our own bodies in the proper ratios, bathes the brain in the chemicals of ecstasy.

GETTING HIGH ON YOUR OWN SUPPLY

When I meditate, I can feel the moment when each drug in the cocktail kicks in. First, I use EFT tapping and release any and every negative thought, emotion, and energy.

This drops my level of cortisol, along with suppressing the high beta brain waves of stress. I now have a molecular substrate in my brain upon which I can build a deep and focused meditative experience.

Next, I close my eyes and focus. **Dopamine** kicks in as I anticipate the delicious hormone and neurotransmitter drug cocktail I'm about to be rewarded with. The dopaminergic reward system of my brain fires up and the "body learning" of how to meditate—stored in my basal ganglia, which memorize frequently performed actions—comes online. *Ingredient one.*

My mind starts to wander. My email inbox. The morning's first meeting. The laugh line of the movie I watched last night. An overdue deadline. Damn, I'm way out of the zone already, cortisol rising, and I haven't been meditating more than 5 minutes.

Dopamine brings me back to focus, aided by **norepinephrine**. I'm motivated. I want Bliss Brain more than I want an endless loop of the Me Show. I return to center. Cortisol drops. Ahhh, I'm back. Norepinephrine stimulates my attention. *Ingredient two.*

Then I realize that my body is uncomfortable. I have a twinge in my right knee. My lower back hurts. My tummy's rumbling because it's empty. I consciously shift my wandering mind back into focus. Back in sync, my neurons secrete **beta-endorphin**, which masks the pain. The discomfort drops away, and being in a body feels wonderful. *Ingredient three.*

I tune in to each of the archetypal strands that guide me. Mother Mary. Kwan Yin. Healing. Strength. Beauty. Wisdom. I imagine myself meditating in a field of a million saints. I'm lost in Bliss Brain, as **serotonin**, the satisfaction drug, kicks in. *Ingredient four.*

I feel one with the universe. **Oxytocin** starts to flow, as I bond with everything. *Ingredient five.* That releases **nitric oxide** and **anandamide**. *Ingredients six and seven.* I float out of my body on a cloud of bliss.

I see stars and lights and hear music, as the psilocybin-like properties of serotonin trigger hallucinations in my brain. I've now whipped up the whole ecstatic cocktail of SONDANoBe in my head.

My whole body begins shaking in ecstasy, as yet more anandamide floods my synapses. Not for nothing were the 19th-century religious mystics called "Shakers" or "Quakers." I relish the orgasmic ecstasy of Bliss Brain, as the SONDANoBe cocktail floods my body and brain. I stay there for 20 or 30 minutes, sometimes an hour.

5.19. Ecstatic mystical experiences are common throughout history. In 1601, Renaissance painter Giovanni Baglione depicted "The Ecstasy of St. Francis." See full-color version in center of book.

As I prepare to end the meditation, I drop back into my body and open my eyes. SONDANoBe has transformed my brain and with it, my perceptions. The world seems exquisite. The air shimmers with aliveness. Colors glisten with intensity. The world seems filled with an endless supply of miracles. The day ahead feels like a gift from the cosmos. I am as excited as a child opening a toy box.

My heart is bursting with gratitude. For the glittering air I breathe. For the miracle of having a human body. For being able to walk, run, laugh, and play. For the astonishing planet on which I live. For the mystery of being one with the universe while at the same time being one with my precious community of fellow beings.

How can anyone be so blessed? The magnitude of grace is bigger than my heart and mind can hold, and it overflows my capacity to receive. Tears of stunned gratitude run down my cheeks. Prolactin unleashes waves of physical satisfaction through my body.

I prepare to meet the day ahead, the chemicals of ecstasy still flowing through my body. I've received my daily "fix" of SONDANoBe, and I go into the day feeling wonderful.

In Bliss Brain, your pleasure-inducing molecules are not dancing alone in your brain and body. They are all dancing together in a glorious synthesis of molecular harmony.

Tears of Bliss Running Down My Cheeks

by Toni Tombleson

"All I can say is WOW! I have been meditating often for the last year and have NEVER experienced what I just did while following the audio for EcoMeditation. I spend probably 99% of my day full of anxiety, frustration, fear, and anger—just overall burnt out with parenthood and life. I was doubtful I'd have any type of enlightened experience and even kept hearing in my head, 'You're just wasting your time, you can't get to that place inside of you, you've tried a million times before.'

"But when I sent the beam of love from my heart, I instantly started laughing and tears of pure bliss came out of my eyes. It was absolutely amazing and I will be doing EcoMeditation every day. Feeling that love and bliss is my dream. Thank you again!!!!!"

I was so delighted when Toni sent in her story after my book *Mind to Matter* inspired her. And when Juliane, whose story appears below, talked to me after a live workshop. So many people like them have found themselves unable to establish a regular meditation practice.

When you make it easy, as EcoMeditation does, people feel the changes in their bodies. They realize that Bliss Brain is accessible. They feel the rush of pleasure chemicals and feel good, while understanding that they have the power to re-create that state every day. Now they understand that they can get high on the chemicals they synthesize in their own brains and bodies.

You don't need any exogenous source—you can get high on SONDANoBe any time you want and never have a hangover.

SONDANoBe is addictive. You don't need to make a New Year's resolution that you'll meditate every day. You don't need to set up a reminder on your watch or mark your calendar. You don't have to swear a solemn oath or find an accountability partner.

You want nothing more than to reexperience that euphoric state every day. You're as hooked as the heroin addict, but you're addicted to a practice that brings you nothing but benefit.

90 Days of Anandamide

by Juliane Robins-Bass

I've used your EcoMeditation tracks on the Insight Timer app for 90 days straight now. It's changed everything.

Before that, whenever I tried meditating all I could get was frustrated. I closed my eyes and all the problems I was dealing with flashed through my mind. I tried mantras, I tried loving-kindness, I tried watching my thoughts.

That one was easy, except there were jumbles of them. They just jumped from one thing to another, mostly the problems I have in my life. They made Monkey Mind look calm. I read books and went to classes, but I could not make myself do it regularly.

The first time I tried EcoMeditation, it worked. I listened to the guided track and I was there in 5 minutes. I felt sooooo good!

So I decided I would do it every single day for the next 90 days and not miss a single day. I listen along to your voice each morning and sometimes again in the night.

I can't believe the person I'm becoming. I'm much less stressed dealing with my patients. I'm nicer to my hubby and kids. My gut troubles have calmed down. I'm even OK in traffic, which used to give me panic attacks.

I have some regrets about all the years I wasted trying other stuff and wish I had this when I first started college and began getting anxiety. But I have it now, and I'm going to keep using it, like, forever.

TAKING THE COCKTAIL ON THE ROAD

I think I've presented a persuasive case that Bliss Brain feels good—seriously good. But what is its relevance to everyday life? A skeptic at one of my keynote speeches queried, "This all sounds selfish. These people escape the dog-eat-dog world of everyday reality by disappearing into cloud-cuckoo-land. That's nice for them, but how does a blissed-out person contribute to society?"

Studies show: a lot. Bliss Brain produces *massive* increases in productivity, creativity, and problem-solving ability.

Bliss Brain means far more than 1 hour of ecstasy for the meditator. The effects extend to the remaining 23 hours of everyday life, as we first saw in Chapter 4. Researchers using fMRIs find that bathing the brain in SONDANoBe conditions its functioning for the rest of the day. Meditators aren't just more peaceful, focused, and creative *during meditation*. They're more creative, productive, and resourceful *in their daily lives*.

Deep effortless alpha meditative states produce "flow" or being in "the zone." In flow, the prefrontal cortex goes offline, the dance of the molecules begins, and creativity is heightened as a result.

At the Columbia University bioengineering lab, researchers use EEGs to provide feedback to users engaged in a demanding task: flying a plane in virtual reality. They find that emotional arousal affects performance. When we're calm, with the dorsolateral PFC of Chapter 3 calming the limbic system, we perform better than when we're agitated.

Commenting on one of their studies published in the *Proceedings of the National Academy of Sciences*, lead author Paul Sajda, professor of biomedical engineering, observed that, "The whole question of how you can get into the zone, whether you're a baseball hitter or a stock trader or a fighter pilot, has always been an intriguing one . . . we can use feedback generated from our own brain activity to shift our arousal state in ways that significantly improve our performance in difficult tasks—so we can hit that home run or land on a carrier deck without crashing."

5.20. Being in the "zone" of flow dramatically improves performance.

The international consulting company McKinsey performed a decade-long study. They tested high-performance executives called upon to solve difficult strategic problems. In flow states, participants' ability to do this increased by 500%. That's a *fivefold* increase. You can enter flow through meditation, or through high-performance activities. Steven Kotler of the Flow Genome Project says that "neurobiologically, they're the exact same state."

Another study, by the Defense Advanced Research Projects Agency (DARPA), used neurofeedback to induce flow states in warriors. Their ability to solve complex problems and master new skills shot up by 490%. A test at the University of Sydney found that students solved a conceptual problem *eight times better* when they were in flow.

This means that your dance of bliss can continue long after you emerge from your morning meditation. SONDANoBe can infuse your day with well-being and a heightened ability to arrive at creative solutions to any challenges that arise. Harvard professor Teresa Amabile, PhD, found that people are more creative in flow, not just at the time of the flow state, but *as long as 24 hours* after they experienced it.

Tackling difficult problems requires holding many ideas at once, and not being rigidly attached to any of them, as we saw in Chapter 3. Some may even be mutually exclusive. When we gain distance from our local minds in meditation, this opens up perceptual space. People in flow states can consider many options.

Kotler notes that this "knocks out the filters we normally apply to incoming information" and loosens up our identification with a single fixed reality. This greatly expands the range of possibilities our minds can juggle, opening up our creativity and productivity. Meditation produces a high-performance brain, able to solve wicked problems, as we'll discover in Chapter 8.

Take a deep breath, and think for a moment about your life. Imagine being 500% more able to solve knotty problems. Picture yourself being 490% better at acquiring new skills and eight times better at conceptual tasks. That's mental superpower!

What might your health, your work, your love life, and your finances look like if you had that superpower? Probably a whole lot better than they do now. These numbers alone are more than enough reason to cultivate a daily meditation practice.

JUST LIKE DRUGS

"Psychotropic" is the term for drugs that alter your mental state. We've seen how they work by docking with the same receptor sites as the body's endogenous pleasure molecules.

There are several classes of psychotropics, each stimulating a particular type of receptor site. The psychedelics LSD, psilocybin, and dimethyltryptamine (DMT) activate serotonin and norepinephrine receptors. The cannabinoids in marijuana dock with anandamide receptors.

The hallucinogen ayahuasca, like LSD and psilocybin, activates serotonin receptors. It is, however, a combination of two plants that have different properties. The leaves of the shrub *Psychotria viridis* actually contain DMT while the bark of the tropical vine *Banisteriopsis caapi* is used for its monoamine oxidase inhibitors (MAOIs), which prevent the breakdown of DMT in the stomach, allowing the DMT to be absorbed into the bloodstream, travel to the brain, and produce hallucinations.

MAO is an enzyme that breaks down norepinephrine, serotonin, and dopamine. The class of antidepressant prescription drugs known as MAO inhibitors or MAOIs prevent this breakdown from occurring, just as the MAOIs in ayahuasca enable it to activate serotonin.

Physiologically, the path to bliss is the same whether brought on by drugs, meditation, or another mind-altering experience. Chris King, author of *The Cosmology of Conscious Mental States*, observes that "research exploring states of meditation and religious devotion show that these states fall into a physiological spectrum as clearly as natural and pharmaceutically induced states do."

Hamlet on Hash

Was William Shakespeare high when he wrote his timeless plays and sonnets?

Debate has raged back and forth on this question since a study published in the *South African Journal of Science* raised the possibility. Anthropologist Francis Thackeray, of the University of the Witwatersrand and colleagues used a forensic technique called gas chromatography mass spectrometry to analyze the fragments of 24 tobacco pipes from 17th-century Stratford-upon-Avon, including some from Shakespeare's garden.

The researchers found cannabis in eight samples, nicotine in at least one sample, and Peruvian cocaine in two samples. Professor Thackeray found that, "Four of the pipes with Cannabis came from Shakespeare's garden."

5.21. The "high"-ly creative William Shakespeare.

Thackeray suggests that the works of Shakespeare may hold possible clues to his use of marijuana. For example, in Sonnet 76, we find "invention in a noted weed," which could reference Shakespeare's use of "weed" (cannabis) to enhance his creative writing ("invention").

After Thackeray published further research, *Time* magazine fueled the debate with an article entitled "Scientists Detect Traces of Cannabis on Pipes Found in William Shakespeare's Garden."

SO WHY NOT JUST USE DRUGS?

So if you can get all these wonderful Bliss Brain states using ayahuasca, peyote, marijuana, LSD, ecstasy, OxyContin, and other psychotropic drugs, why not just pop an herb or a tablet every time you want to feel high?

You don't have to be a mathematician to compare the level of effort required of the meditators of Chapter 3 with the drug-poppers of Chapter 5 and reason: "Hmmm . . . 10,000 hours of meditation versus one pill. Why not the pill?"

Two words: side effects.

Chemically induced joy comes at a cost. That cost can be high. Very, very high. So high that you're going to think twice after reading what science has to say about drug use.

One study found that adolescents who smoke just a couple of joints of marijuana show changes in their brains. That's not a couple of years of smoking or the decades that some adults rack up. It's just *two joints.*

A research team led by Dr. Gabriella Gobbi, a professor and psychiatrist at the McGill University Health Center in Montreal, discovered that teenagers using cannabis had a nearly 40% greater risk of depression and a 50% greater risk of suicidal ideation in adulthood.

Dr. Gobbi stated that "given the large number of adolescents who smoke cannabis, the risk in the population becomes very big. About 7% of depression is probably linked to the use of cannabis in adolescence, which translates into more than 400,000 cases."

The research that revealed these startling numbers was not just a single study of adolescent marijuana use. It was a meta-analysis and review of 11 studies with a total of 23,317 teenage subjects followed through young adulthood.

Further, Gobbi's team only reviewed studies that provided information on depression in the subjects prior to their cannabis use. "We considered only studies that controlled for [preexisting] depression," said Dr. Gobbi. "They were not depressed before using marijuana, so they probably weren't using it to self-medicate." Marijuana use *preceded* depression.

The specific findings of Gobbi's research include:

- The risk of depression associated with marijuana use in teens below age 18 is 1.4 times higher than among nonusers.
- The risk of suicidal thoughts is 1.5 times higher.
- The likelihood that teen marijuana users will attempt suicide is 3.46 times greater.

In adults with prolonged marijuana use, the wiring of the brain degrades. Areas affected include the hippocampus (learning and memory), insula (compassion), and prefrontal cortex (executive functions).

The authors of one study stated that "regular cannabis use is associated with gray matter volume reduction in the medial temporal cortex, temporal pole, parahippocampal gyrus, insula, and orbitofrontal cortex; these regions are rich in cannabinoid CB1 receptors and functionally associated with motivational, emotional, and affective processing. Furthermore, these changes correlate with the frequency of cannabis use . . . [while the] . . . age of onset of drug use also influences the magnitude of these changes."

A large number of studies show that cannabis use both increases anxiety and depression and leads to worse health. Key parts of your brain shrink more, based on how early you began smoking weed, and how often you smoke it. That's a "high" price to pay.

How about MDMA, aka ecstasy? Is that a shortcut to Bliss Brain? Twenty-five years of research into MDMA have shown that it has serious side effects, including impairments to cognitive function, sleep, memory, social intelligence, problem-solving ability, and mental health. It is capable of increasing cortisol, the main aging hormone, by 800%. A research review concludes that, "The damaging effects of Ecstasy/MDMA are far more widespread than was realized a few years ago, with new neuropsychobiological deficits still emerging."

And the side effects of prescription psychotropics aren't pretty either, from sexual dysfunction to mania to dementia.

5.22. Many prescription drugs come with a big dose of side effects.

So I flip the "Why not use drugs?" question on its head.

Why use drugs? Your own body synthesizes the chemicals of ecstasy all by itself. It does this in concentrations that are not harmful, are in perfect balance, have the correct ratios one to another, degrade effortlessly when they are no longer needed, create no hangover, and produce no side effects. When Bliss Brain is achievable daily, consistently, easily, safely, and on demand, why seek exogenous sources of ecstasy?

A Native American medicine woman told me that in her Twisted Hair clan, one of seven that make up the Cherokee nation, teachings about psychoactive herbs or "plant medicine" are passed from generation to generation. She said, "If you take plant medicine, you will have the [enlightenment] experience. But you will not grow as a human being unless you learn to create the experience within yourself."

"Medicines" that open you to nature's deepest truths can be a powerful ally in your personal evolution; dependence on those medicines to reach your most valued states of consciousness can be an alluring trap.

THE POWER OF ADDICTION

Long before I met Christine and got married, I lived alone but enjoyed renting rooms to interesting housemates.

Brandon was the tenant who lived with me the longest. As the years went on, I grew to admire him greatly. A minister ordained in the Unity Church, he also ran the local Alcoholics Anonymous (AA) chapter. He'd been an alcoholic until 15 years earlier, when he'd wrecked his car and broken his back in a catastrophic drunken crash. He went on to "hit bottom" in every way till becoming sober through AA.

I went with him to AA meetings several times. The stories I heard there astounded me. I met one of Brandon's regular attendees named Greg. Here is his story.

Ignoring Stop Signs

I work construction. From when I was a teenager. Our crew used to hang out after work and drink beer. We also began drinking during our lunch break.

It didn't affect me at first. I met Marguerite and we got hitched. We had our two kids, Sammie and Cass. We were doing OK, though we used to fight a lot.

When the kids were in grade school, I wasn't just drinking at lunch and going to the bar on the way home. I was drinking before I went to work. I used to buy a suitcase of Bud on the way to the construction site.

One day, my friend Job, who like me was plastered, cut off his finger with a table saw. He was just so drunk he couldn't see straight.

That didn't stop me. I was doing weed most days, coke on the weekends, as well as the suitcase of Bud every morning.

Marguerite said she'd leave me if I didn't stop drinking. Our finances were going south, even though I was making plenty of money. I was spending an ungodly amount. The coke especially was expensive.

We lost the house because I wasn't making the payments. We went to live in a trailer in a skanky trailer park. My credit cards were maxed out and I declared bankruptcy.

Eventually, Marguerite couldn't take it anymore and moved out with the kids. She filed for divorce. The judge said I couldn't see the kids unless I stopped using. That really hurt, because I loved those kids.

But I couldn't stop taking the stuff, and the judge put a restraining order on my ass and I couldn't see my kids anymore. That hurt like hell. The only thing that helped me forget was using.

My boss told me I'd lose my job if I came in drunk one more day. They sent me to rehab, but it didn't help. I can't believe they didn't terminate me sooner.

I was living on the street, no job, no wife, no kids, no house. I spent any money I had on drugs. I weighed close to 100 pounds because I wasn't eating.

One day some guys beat me up and shook me down for the little bit of weed I had. In the hospital, the guy in the bed next door told me about AA. He took me to a meeting and became my sponsor. I got sober.

I've now been sober for 7 years, 6 months, and 1 day. I'm an alcoholic and I always will be. But I've rebuilt my life. I'm a sales manager for a mobile home company. I get to see my kids under court supervision. I'm dating a nice woman. I look back at the asshole I used to be and I can't believe it's the same person. Thank God.

I was astonished as I listened to Greg's story because there were so many danger signs along his path to the bottom. Each screamed "Stop" ever louder. The fights with his wife. The gruesome accident with his workmate's finger. The escalating drinking—how many cans are a "suitcase" of beer anyway? The highest most people can count is a six-pack.

His wife's threats to leave. Escalating to harder drugs. Financial struggles. Bankruptcy. The trailer park. His wife taking the kids. The restraining order.

The red line that he would get fired. Rehab. Getting fired. Living on the street. Getting beat up.

Greg ignored all those signals. That's how strongly the power of addiction held him. Until I went to AA and heard the stories that Greg and other addicts told, addiction was just an idea to me. I had no idea of the power that drugs could exert over a person's behavior, and how impossible it is for addicts to stop even when they have everything to lose.

HAPPINESS ADDICTION

Think about the strength of Greg's addiction. It was more powerful than his need for love with his wife and children. It was more powerful than his need for money. For a job. For security. For home. For a roof over his head. For his health. His addiction was bigger than every other part of his life. That's how powerful a pull toward a chemical can be.

And that's how powerful this cocktail of seven magic endogenous chemicals can be—except that they're good for you. Once you've tasted the elixir and then got in the habit of sampling SONDANoBe daily, it becomes as addictive as a street or prescription drug. Move over, OxyContin.

You wake up in the morning and the first thing you think of is Bliss Brain. Getting your "fix" of anandamide, oxytocin, norepinephrine, nitric oxide, serotonin, dopamine, and beta-endorphin is your first priority.

You're focused on feeling good—from the inside out—from the moment you open your eyes. You crave your fix of SONDANoBe. You throw on some clothes and make tea or coffee. Quickly, so you can enter that magical state as soon as possible.

Sitting in your favorite position and closing your eyes, you tiptoe over the threshold of experience and into the mystery. You drop into the heart of the universe. You're there. The cascade of SONDANoBe floods your brain. You're hooked, drawn up into the light.

When you emerge from meditation, you're more compassionate, emotionally balanced, mentally coherent, effective, kind, creative, healthy, and productive. The effects ripple through the whole community around you. At the center of that circle is a great-feeling you.

The Gregs of this world go for heroin, weed, or alcohol to make themselves feel good. That's simply because they don't realize that a far better drug is available. *SONDANoBe is what addicts are really craving.* They want to feel good, but they're looking for exogenous chemicals to meet their needs. They don't understand that what they're searching for is right inside their own brains. The only reason those drugs feel good to the Gregs of this world is that they're facsimiles of the substances that their own brains produce.

Bliss Brain is a formula, just like the World's Best Cocktail. It's the World's Best High, and it's just as addictive. The brain that experiences SONDANoBe once can never go back to its old state. By remodeling neural tissue, SONDANoBe consolidates learning and hardwires bliss.

While street drugs shrink and damage vital brain regions, SONDANoBe does the opposite. It grows your brain. It expands the brain regions that regulate your emotions, synthesize great ideas, stimulate your creativity, acquire new skills, heal your body, extend your longevity, improve your memory, and boost your happiness. The next chapter shows how a brain bathed in the chemicals of ecstasy starts to *change its fundamental structure*, as the software of mind becomes the hardware of brain.

DEEPENING PRACTICES

Here are practices you can do this week to integrate the information in this chapter into your life:

- **Noticing SONDANoBe:** When you meditate, notice the emotional sensations you feel in your brain. Notice as each neurochemical hits; for instance, when you feel the warm rush of connection, know that's oxytocin. Or when you feel alert, recognizing that norepinephrine is being released. This practice helps you understand the contribution of each neurochemical to the meditative experience.
- **Attention Practice:** A common issue for meditators is falling asleep. That rarely happens if you're meditating effectively, because you're producing norepinephrine. This week, notice the sensations in your head when you're extremely focused during a meditation. You might feel a tingling at a particular location inside your skull, or a straightening of your spine. These symptoms of alertness mean that norepinephrine is kicking in. Intensify those sensations, and see how high you can dial them up. If you practice alertness consistently, it becomes stronger.
- **Listening Touch:** This is a practice to do with a partner. Set a timer for 15 minutes. Sit back-to-back with your partner and close your eyes. Touch their neck, shoulders, and hands, tuning in to their mood. When the timer ends, switch roles. The giver becomes the receiver and vice versa. Also experiment with doing this face-to-face, but only after you've done it back-to-back a few times. In the study, doing this three times per week produced oxytocin surges and lowered blood pressure in men.
- **Connecting with Other Species:** If you have a dog, you can do this easily. If you don't, find a friendly one in a public place, or do it with a friend's dog with whom you have a relationship. Gaze gently into the dog's eyes, while feeling centered in your heart. Breathe from your heart. Notice how your body feels, and measure for how long the practice feels natural. Don't try this with cats; for the feline brain a stare is a symbol of aggression. Sorry again, cat people.

EXTENDED PLAY RESOURCES

The Extended Play version of this chapter includes:

- Dawson Guided EcoMeditation: Becoming One with the Universe
- Love Is Lifting Me Higher Chant with Julia and Tyler

- Flow States and Creativity
- Side Effects of Marijuana and Ecstasy
- Dog Gazing and Oxytocin

Get the extended play resources at: BlissBrainBook.com/5.

CHAPTER 6

CHANGING THE HARDWARE OF BRAIN WITH THE SOFTWARE OF MIND

Science divides the study of organisms into two basic categories: anatomy and physiology. Anatomy describes *structure* while physiology describes *function*.

Your digestive tract, for instance, starts with your mouth, continues with your esophagus, extends to your stomach, and goes on from there. That's anatomy. It names component parts and describes their structure.

But if you were a silicon-based alien who lived off cosmic radiation, you would have no concept of a human digestive tract. You could look at an X-ray of the anatomy of the digestive system, clearly seeing the structure of every part, yet have no idea what the system does or the purpose it serves. It is like you or me looking at an ancient Mayan calendar stone: We observe every detail of its structure, but we have no idea how it works.

That's where physiology comes in: It shows us how systems function. The mouth chews and breaks up food. The esophagus transports it to the stomach. The stomach contains digestive juices that break down the chunks into nutrients that can be absorbed later in the intestines. When you understand the physiology of a system, the anatomy makes sense. You need to understand both form (anatomy) and function (physiology) for a complete picture of a biological system.

TURNING PHYSIOLOGY INTO ANATOMY

From the dawn of recorded human history, people have known that physiology can *create* anatomy. Ancient Spartans put warriors through a rigorous and structured physical development course to grow their muscles and enhance their martial abilities. It included activities such as rock lifting, stone throwing, wrestling, and rope climbing.

They understood that if you work a muscle hard (physiology), it grows (anatomy).

6.1. The idea that physiology (exercise) can become anatomy (big muscles) has been understood for millennia.

The same applies to many other types of training. As fighter pilots progress through rigorous flight training, their reaction time shortens. An opera singer might not be able to hit high C at her first attempt, but with practice she can sing it on pitch every time.

While the principle that exercising a body system physiologically produces changes in anatomy has been uncontroversial in other disciplines, many scientists dismissed the idea when it came to neurons.

Neural plasticity was suggested by advanced thinkers in the 1930s and demonstrated experimentally in the 1960s, yet was met with skeptical denial by the neuroscience establishment.

Unbelievably Changing Neurons

In 1964, neuroscientist Marian Diamond (1926–2017) published research that changed the field of neuroscience forever. Hers was the first study to demonstrate brain neuroplasticity. She showed that the brain can grow as a result of experience.

These findings were in direct opposition to scientific thinking at the time, which held that brain change happened in only one direction: deterioration with age.

In the groundbreaking experiment, Dr. Diamond and colleagues David Krech and Mark Rosenzweig studied rats in an enriched environment, with cage mates and toys. The control rats were in an impoverished environment.

The result?

The cerebral cortex of the enriched rats was 6% thicker than that of the impoverished rats. "This was the first time anyone had ever seen a structural change in an animal's brain based on different kinds of early life experiences," observed Dr. Diamond.

In 1965, she presented her research at the annual meeting of the American Association for Anatomy. She recalls being scared as she faced a huge room containing few female faces. After her presentation, "a man stood up in the back of the room and said in a loud voice, 'Young lady, that brain cannot change!'. . . I simply replied, 'I'm sorry, sir, but we have the initial experiment and the replication experiment that shows it can.'"

She notes that "it took two generations of researchers for the 'enrichment paradigm' to be validated."

In 1985, Diamond demonstrated that the brains of older mammals could grow too. After just 6 months in a stimulating environment (physiology), the cortex of older rats—the rat equivalent of a 75-year-old human—was thicker (anatomy). Diamond observed that, "This means you have some degree of *control* over your own brain tissue."

Diamond lists five factors that are important for a healthy brain:

1. Diet
2. Exercise
3. Challenge
4. Newness
5. Love

She often said that the phrase "Use It or Lose It" applies to the brain as well as to muscles. Diamond walked her talk, continuing to research and teach until just a couple of years before her death at age 90.

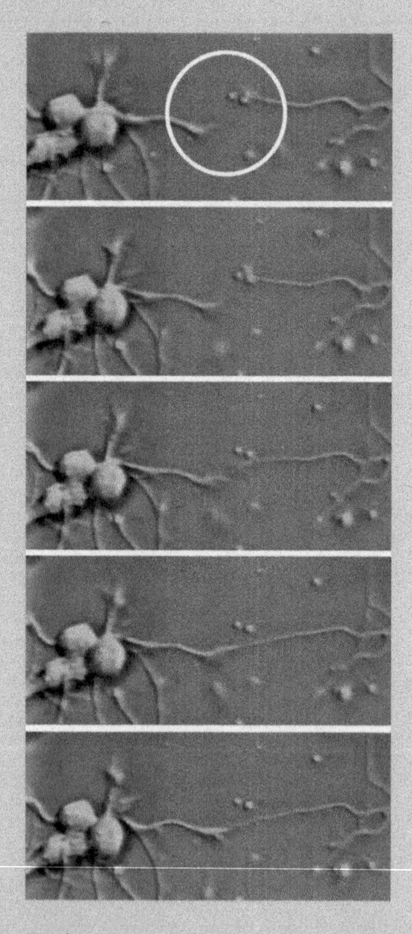

6.2. Scanning electron microscope images showing how fast two neurons wire together. Top: beginning of stimulation. Middle above: after 7 seconds. Middle center: after 8 seconds. Middle below: after 11 seconds. Bottom: after 12 seconds.

THE TRANSLATIONAL GAP

The popular image of a scientist is a disinterested and objective observer who dispassionately studies empirical data.

But in reality, science is marked by fads, trends, paradigms, fashions, feuds, warring camps, petty jealousies, and die-hard beliefs. Conventional science usually reacts to new findings with disparagement. When confronted with the evidence for energy healing, one skeptic exclaimed, "I wouldn't believe it, even if it were true!" Innovation faces daunting headwinds.

The opposition to new therapies has unfortunate side effects. A group of distinguished colleagues and I analyzed US government reports on health-care innovation. We found that the average medical breakthrough takes 17 years to get from lab to patient.

Even more startling, only 20% of new treatments jump this "translational gap." The other 80% are lost forever. The result is that when we seek treatment, *we are getting only one fifth of 17-year-old medicine.*

We would be outraged if we were forced to use a cell phone that was 17 years old, with 80% of its features disabled. But as a society, we treat this paradigm as perfectly reasonable when it comes to taking care of our precious and irreplaceable bodies.

The neuroscience establishment fought the idea of neural plasticity tooth and nail. Yet eventually the evidence became too overwhelming to deny, and the weight of scientific opinion began to change.

The rats that Marian Diamond studied had either an enriched or an impoverished environment. That changed their brain state. If you're surrounded by a nurturing physical, emotional, mental, and spiritual environment, you're in one brain state. If you're surrounded by danger, uncertainty, and hostility, you're in a quite different brain state.

Brain states, along with mental, emotional, and spiritual states, run the gamut. When the brain's Enlightenment Circuit is turned on, you're in a happy and positive state. When the Default Mode Network (DMN) of Chapter 2 predominates, you're in a negative and stressed state.

State Progression

Cognitive psychologist Michael Hall has been fascinated by human potential for over 40 years. He has studied the most advanced methods, authored more than 30 books on the topic, and mapped the stages by which people change.

Unpleasant experiences are what usually motivate us to change. These involve mental, emotional, or spiritual states. Examples of such states are despair, stagnation, anger, or resentment. Hall calls these "unresourceful" states.

We can cultivate resourceful states, such as joy, empowerment, mastery, and contentment. To describe the movement of a person from an unresourceful to a resourceful state, Hall uses the term "state progression."

Hall's "state progression" model has several steps:

- Identify the unresourceful state.
- Identify the desired state.
- Countercondition dysfunctional behavioral patterns that maintain the unresourceful state.
- Activate change toward the desired state.
- Experience the target state.
- Repeat the experience of the desired state.
- Condition new behaviors that reinforce the desired state.

That's the promise of directing your attention consciously rather than defaulting to the brain's negativity bias. Attention sustained over time produces state progression and triggers neural plasticity. If you focus on positive beliefs and thoughts repeatedly, bringing your mind and focus back to the good, you then use attention in the service of positive neural plasticity.

When we have practiced sufficiently to be able to maintain this focus, we achieve a condition that Hall calls *positive state stability.* Our minds become stable in that new state. Their default setting is no longer to focus on the negative. The brain's negativity bias is no longer hijacking our attention and directing it toward the negative things that are happening, either in our own lives or in the world. We have moved through the stages of state progression to positive state stability.

ALTERED TRAITS

Science journalist Daniel Goleman and neuroscientist Richard Davidson have been friends since college. They teamed up to write a genius book called *Altered Traits* that reviews the extensive body of fMRI research on the effect of meditation on the brain.

Why do they use the word "traits" and not simply "states"?

In Chapter 3, we examined which brain regions are activated by meditation. The meditator with those brain regions active experiences a "state." This is analogous to a "state" of bliss or a "state" of elation or a "state" of surprise. Or under the sway of the DMN, a "state" of anger, fear, disgust, or despair.

States are transient. They come and go.

Traits, on the other hand, are hardwired brain circuits that control behavior. Psychologists can measure a trait like conscientiousness, extroversion, kindness, confidence, or compassion.

A trait is a characteristic of a person, a norm to which they revert. We've all known people with traits like pessimism, selfishness, or impatience—and they're less fun to be around.

We've known other people with traits like patience, understanding, and empathy, and our experience of such people is very different. Once we know someone has a trait, we can count on them to display it routinely.

The title of Goleman and Davidson's book *Altered Traits* refers to the way meditation changes our personalities. As we experience positive psychospiritual *states,* we achieve positive state stability. We "cultivate enduring qualities like selflessness, equanimity, a loving presence, and impartial compassion—highly positive altered *traits.*"

Meditation activates the Enlightenment Circuit we examined in Chapter 3. It triggers the experience of altered *states* of consciousness. When we achieve positive state stability, neural plasticity kicks in, stimulating the growth of neural pathways to create new *traits.*

6.3. State anxiety. A woman is anxious having just read the newspaper.

6.4. Trait anxiety. She has an anxious outlook on everything.

Andrew Newberg, co-author of *How Enlightenment Changes Your Brain,* observes that the brain that has an enlightenment experience "is not the same brain . . . subtle and permanent changes will have taken place in key areas."

In the One Percenters of Chapter 4, *the physiology of altered states* becomes *the anatomy of altered traits*. The neural wiring in happiness-producing areas of the brain becomes bigger and faster, while the wiring in the misery networks shrinks and slows.

BICYCLES AND HABITS: MOVING BEHAVIOR FROM THE TOP TO THE BOTTOM OF THE BRAIN

At first, you have to focus intently when learning a new skill. Think about the first time you rode a bicycle. You focused intently on balancing. This lights up the executive regions of the brain's PFC, the top level of neocortex function. The *top of the brain* is working hard to provide the focus to execute the skill.

You were also afraid of falling, combined with exhilaration as you picked up speed. These emotions engaged the brain's reward circuits to reinforce your learning.

But as you became proficient, riding a bike took less focus. Eventually you could pick up a bike and balance without any conscious thought at all. Even years later, you retained the skill.

That's because as we practice a behavior repeatedly, we grow neural networks in the *bottom of the brain.* Structures like the basal ganglia in the striatum take over execution of a familiar task. These new neurons make the task reflexive so it runs below the level of conscious awareness. Bottom-up automation takes over from intensive, focused, top-down control.

The same learning and automation process occurs with any behavior we practice consistently, including meditation. Meditation initially requires top-down PFC control over selfing and attention. Positive states then activate the brain's reward circuits.

Dopamine kicks in to reinforce the pleasant moods the brain is learning. This transfers them from the top levels of the PFC (where they require willpower) to the basal ganglia (where they're an automatic reflex). This "reconsolidation" reinforces positive state stability to produce positive traits.

One team of neuroscientists notes that "with appropriate training and effort, people can systematically alter neural circuitry associated with a variety of mental and physical states that are frankly pathological." We can take our dysfunctional brain anatomy and rewire it for happiness and creativity. Bliss Brain becomes our new normal, as we'll see in the Afterword.

6.5. Riding a bike was hard at first, but as the neural wiring for the skill grew in the brain's basal ganglia, it became reflexive.

Which Came First: The Chicken or the Egg?

Sara Lazar, PhD, professor of psychology at Harvard Medical School, was one of the first researchers to investigate brain changes associated with meditation. The results of her studies are startling.

It all began when Lazar was working on a doctorate in molecular biology and training for the Boston marathon. Running injuries sent her to a physical therapist who told her she needed to stretch more, so she started going to a yoga class.

"The yoga teacher made all sorts of claims, that yoga would increase your compassion and open your heart. And I'd think, 'Yeah, yeah, yeah, I'm here to stretch,'" Lazar recalls. "But I started noticing that I was calmer. I was better able to handle more difficult situations."

Curious, she ran a literature search and discovered studies detailing the benefits of meditation. She was hooked and thereafter changed her research focus.

Her first study on meditation, published in 2005, investigated the hypothesis that meditation might be linked with actual changes in the brain's structure. The subjects were 20 long-term meditators who meditated an average of 40 minutes a day. They were compared to a control group of 15 non-meditators.

The study found that the long-term meditators had increased gray matter in the insula and the PFC regions associated with sensory processing. Lazar notes that this is logical, as during meditation the focus is on the senses not on thinking.

The greatest thickening of the gray matter was among those who meditated the deepest, as measured by breath rate. "This strongly suggests that the differences in brain structure were caused by the meditation rather than that differences in brain thickness got them into meditation in the first place," observes Lazar.

The study's authors concluded, "These data provide the first structural evidence for experience-dependent cortical plasticity associated with meditation practice."

Lazar's second study, published in 2011, decisively answered the question of *which came first,* the chicken or the egg—meditation or increased brain volume. A control group of 17 non-meditators was compared to 16 who went through an 8-week mindfulness-based stress reduction (MBSR) program. It included a weekly class and 30 to 40 minutes of exercises each day.

The brain volume of the meditators increased in four areas: the PCC (managing mind wandering and selfing), hippocampus (emotions and learning), temporoparietal junction (empathy and compassion), and pons (neurotransmitter producer). Volume decreased in the amygdala (fight-or-flight trigger).

All this after *just 8 weeks* of mindfulness practice!

Sara Lazar has now been meditating for more than 20 years and reports that "it's had a very profound influence on my life. It's very grounding. It's reduced stress. It helps me think more clearly. It's great for interpersonal interactions. I have more empathy and compassion for people."

MAPPING THE REGIONS THAT GROW

Which brain regions change with meditation practice? fMRI machines have been improving in resolution over the past few years, just as the resolution on your laptop has become sharper and richer. This is enabling scientists to build up a progressively more detailed picture of which brain regions are growing and shrinking in the One Percenters of Chapter 4.

These areas are emerging as key regions changed by meditation:

- Amygdala, hippocampus, thalamus, and other structures in the emotional midbrain. Central in stress, relaxation, memory, and learning.
- Anterior cingulate cortex. Involved in controlling the focus of attention.
- Caudate nucleus. Involved in memory storage and processing, the caudate nucleus plays an essential role in how the brain learns, using feedback from past experience to influence current actions.
- Areas of the cingulate cortex responsible for regulating the brain's own activity.
- Insula. Makes us aware of our internal emotional states and body sensations.
- Medial prefrontal cortex. Influences emotional responses in memory and decision-making.
- Orbitofrontal cortex. Involved with rational thought, impulse control, cognitive reasoning, and personality.
- Posterior cingulate cortex. One of the two nodes in the DMN, it's active in memory retrieval and attaching significance to perceptions.
- Prefrontal cortex centerline regions related to paying close attention.
- Somatomotor areas processing pain, touch, and orientation of the body in space.
- Striatum, as well as limbic and prefrontal regions involved in emotional self-control and craving.

We'll look at each of these in turn because understanding their functions will show you how they contribute to your meditation practice. By the end of this chapter, you'll understand each region activated in Bliss Brain, how they integrate into *four distinct networks,* and how these networks *coordinate* to produce elevated states of flow.

The research into brain changes associated with meditation is relatively new. At this early stage of scientific discovery, we don't yet have a complete catalog of meditation's effect on the brain and the picture will keep changing as ever-higher-resolution machines and new technologies are developed.

Also, *Bliss Brain* isn't designed to be a textbook, defining the precise boundaries of current science. It's an overview for the general reader, simplifying enormously complex biological interactions to provide an outline of what's happening in the brain. However, that outline gives us a fascinating picture of the brain regions active in meditation and how they are changed by practice.

That change happens quickly. One study looked at brain scans of elderly people doing a chanting meditation called Kirtan Kriya. They had thicker frontal lobes, and the activity of the thalamus—the emotion regulation structure reviewed in Chapter 3—showed that both right and left hemispheres were acting symmetrically. They had improved concentration and memory.

How long did they have to practice for this big payoff? They meditated for *just 12 minutes a day* for 8 weeks.

6.6. Kirtan Kriya meditation uses specific hand positions.

Size Does Matter

After Einstein died in 1955 at the age of 76, his brain was removed, photographed, and preserved by pathologist Thomas Harvey at Princeton Hospital in New Jersey. A bizarre series of events followed, including the purported theft of the brain, but that's another story.

Many of the photographs went unseen by the public or by researchers until they were bequeathed by Dr. Harvey's estate to the National Museum of Health and Medicine in Washington, DC, in 2010.

A study published in 2014 analyzed the newly unearthed photographs and determined that Einstein had an "extraordinary" prefrontal cortex, and that other areas of his brain were larger than those areas not only in men of his age but in young men as well.

The researchers, Florida State University anthropologist Dean Falk along with physicists from East China Normal University in Shanghai, found that Einstein's corpus callosum, in particular, was enlarged in comparison to those of 15 elderly healthy males and 52 young controls.

As we saw in Chapter 3, the corpus callosum connects the right and left hemispheres of the brain, allowing them to communicate and coordinate. Earlier research by Falk and others had shown that sections of Einstein's parietal lobe were larger by 15%.

One part of the corpus callosum that showed enlargement was the splenium, which is responsible for making connections among the occipital, parietal, and temporal lobes as well as between those lobes and the prefrontal cortex.

Dr. Falk and his researchers concluded: "These findings show that the connectivity between the two hemispheres was generally enhanced in Einstein compared with controls. The results of our study suggest that Einstein's intellectual gifts were not only related to specializations of cortical folding and cytoarchitecture in certain brain regions, but also involved coordinated communication between the cerebral hemispheres."

THE FOUR KEY NETWORKS

Cataloging the brain regions changed by meditation is fascinating, but even more interesting is *how they work together.* Brain regions work together in *networks of activity,* and a useful perspective from which to view them is the *function* of each network.

There are four main neural networks transformed by meditation, and Goleman and Davidson define their functions as: Emotion Regulation, Attention, Selfing, and Empathy. In combination, these networks function together in what Newberg calls the "Enlightenment Circuit."

We'll explore each of the networks and identify the role of the brain regions identified earlier in the chapter in the functioning of that network.

THE EMOTION REGULATION NETWORK

First comes the Emotion Regulation Network. I consider this primary, because I believe that unless we have the ability to regulate our emotions, we cannot enjoy a happy life. We can't sustain Bliss Brain for long enough to spark neural plasticity if our consciousness is easily hijacked by negative emotions like anger, resentment, guilt, fear, and shame. The Emotion Regulation Network controls our reactivity to disturbing events.

Regulating emotions is the meditator's top priority. Emotion will distract us from our path every time. Love and fear are fabulous for survival because of their evolutionary role in keeping us safe. Love kept us bonded to others of our species, which gave us strength in numbers. Fear made us wary of potential threats. But to the meditator seeking inner peace, emotion = distraction.

In the stories of Buddha and Jesus in Chapter 2, we saw how they were tempted by both the love of gain and the fear of loss. Only when they held their emotions steady, refusing either type of bait, were they able to break through to enlightenment.

THE HOSTILE TAKEOVER OF CONSCIOUSNESS BY EMOTION

Remember a time when you swore you'd act rationally but didn't? Perhaps you were annoyed by a relationship partner's habit. Or a team member's attitude. Or a child's behavior? You screamed and yelled in response. Or perhaps you didn't but wanted to.

So you decided that next time you would stay calm and have a rational discussion. But as the emotional temperature of the conversation increased, you found yourself screaming and yelling again. Despite your best intentions, emotion overwhelmed you.

Without training, when negative emotions arise, our capacity for rational thought is eclipsed. Neuroscientist Joseph LeDoux calls this "the hostile takeover of consciousness by emotion." Consciousness is hijacked by the emotions generated by fearful unwanted experiences or attractive desired ones. We need to regulate our emotions over and over again to gradually establish positive state stability.

In positive state stability, when someone around us—whether a colleague, spouse, child, parent, politician, blogger, newscaster, or corporate spokesperson—says or does something that triggers negative emotions, we remain neutral.

The same applies to negative thoughts arising from within our own consciousness. Positive state stability allows us to feel happy despite the chatter of our own minds.

Getting triggered happens quickly. LeDoux found that it takes less than 1 second from hearing an emotionally triggering word to a reaction in the brain's limbic system, the part that processes emotion. When we're overwhelmed by emotion, rational thinking, sound judgment, memory, and objective evaluation disappear.

But once we're stable in that positive state, we've inoculated ourselves against negative influences, both from our own consciousness and from the outside world. We maintain that positive state over time, and state becomes trait.

6.7. Once the Emotion Regulation Network is developed, we're no longer triggered by emotional responses to the events around us.

Hippocampus

Emotional self-regulation isn't just "emotional" in the sense of feelings, such as anger, shame, guilt, and resentment. It's physical, in the form of bundles of neurons that fire together and wire together, sometimes communicating with distant parts of the brain. Behavior that demonstrates poor emotional self-regulation is the external evidence of the activity of neural pathways deep inside the limbic system.

In people who are depressed, the hippocampus shrinks over time. In people with chronic PTSD, high cortisol levels produce calcium deposits on the hippocampus. You want lots of calcium in your bones and teeth. You certainly don't want it ossifying your brain's memory and learning center.

Conversely, people with effective emotional self-regulation grow a larger hippocampus and much greater volumes of neural tissue in substructures like the dentate gyrus, a part of the hippocampus that coordinates emotional control among different parts of the brain. As happy people practice the emotional regulation required to shift their focus away from random thoughts and the problems of life, they turn states to traits.

They also improve their memories. A Harvard study found that meditation leads to increases in the concentration of gray matter in the hippocampus. When you activate the hippocampus by regulating your emotions, you build the hardware for both happiness and learning.

Thalamus

Big bundles of neurons conduct information up through the spinal cord into the brain. Sitting at the top of this conduit is the thalamus. In Chapter 3, I compared the thalamus to a relay station, conducting information from the senses to the prefrontal cortex.

During meditation, the thalamus is active, as meditators suppress sensory input that might pull them out of Bliss Brain. Andrew Newberg finds that one of the two lobes of the thalamus is often more active than the other.

One interpretation of this activity may be the meditator's awareness that she is more than her body and that she is connected to nonlocal mind, not just her senses. It's the thalamus that is telling us what is and isn't reality, and this is affected by the larger reality in which the meditator is absorbed.

In long-term meditators, this asymmetry persists when they open their eyes. As the meditator experiences oneness, the universe will, in Newberg's words, "be sensed as real. But it will not be a 'symmetrical' reality. Instead, it will be perceived 'asymmetrically,' meaning that the reality will appear different from one's normal perception."

The nonlocal universe may be perceived as more real than local sensory reality and, as Newberg observes, "The more frequently a person engages in meditative self-reflection, the more these reality centers [like the thalamus] change."

The Emotion Regulation Network is the network in which brain changes may happen fastest. In my book *Mind to Matter,* I tell the story of Graham Phillips, an Australian TV reporter who decided to go on an 8-week mindfulness journey. Here's what happened to his brain.

Eight Weeks to a Remodeled Brain

Australian astrophysicist and TV journalist Graham Phillips, PhD, was skeptical about meditation, but the growing research about its benefits persuaded him to embark on a personal test.

"I'd never really contemplated whether meditation could do anything for me," he said, "But the more I hear about the research, the more keen I am to see whether it has any effect. So I'm going to try it myself for 2 months. . . . For me, to take meditation seriously, I need some hard evidence that it's changing my brain for the better."

To that end, Phillips engaged a Monash University team—headed by biological psychology professor Neil Bailey, PhD, and clinical psychologist Richard Chambers, PhD—to run a series of tests on Phillips to evaluate his memory, reaction time, and ability to focus. With MRI scans, they also measured the volume of each region of his brain, especially the areas involved in memory and learning, motor control, and emotion regulation.

Within 2 weeks of starting his new routine, Phillips felt better able to cope with job and life challenges. After 8 weeks he reported that he "notices stress but doesn't get sucked into it."

The objective results, as demonstrated by the second round of tests, were that Phillips was better at behavioral tasks, even though his brain showed less activity. The researchers explained that his brain had become more efficient, doing a better job while expending less energy.

His memory tests improved. His reaction time was cut by almost half a second. That's the difference between life and death when a pedestrian steps in front of your car.

The researchers were particularly interested in his dentate gyrus, since it helps regulate emotion in other parts of the brain and exert control over the unruly DMN. The volume of nerve cells in his dentate gyrus had increased by an astonishing 22.8%.

This change in Phillips's brain indicated a dramatically increased ability to regulate emotions. Psychological tests demonstrated that Phillips's cognitive abilities had increased by several orders of magnitude as well.

A skeptic no more? At the end of his meditation test, Phillips said, "Hey, I'm impressed and certainly feeling happier with my younger, heavier brain."

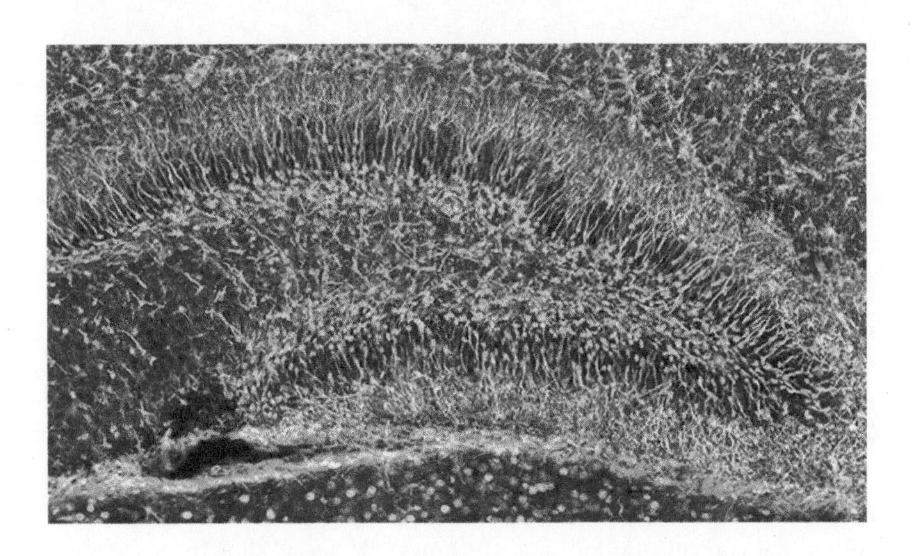

6.8. Neural stem cells in the C-shaped dentate gyrus.
See full-color version in center of book.

Here are the brain regions that are enlarged in the Emotion Regulation Network.

- Ventromedial prefrontal cortex
- Thalamus
- Temporoparietal junction (TPJ)
- Hippocampus
- Subiculum, the main output region of the hippocampus
- Anterior and mid cingulate cortices
- Orbitofrontal cortex
- Precuneus
- Ventromedial orbitofrontal cortex

This last structure exerts control over the amygdala, striatum, and hypothalamus. According to the imaging laboratory team at the University of Hong Kong, it "generates a more flexible assessment to counterbalance the automatic reactivity in amygdala and striatum and has hence an important role in top-down emotional regulation and reappraisal of negative emotional states."

Amygdala

In the One Percenters, some brain regions shrink. The amygdala, the brain's "fire alarm," decreases in volume. In adepts, disuse causes it to atrophy.

The control circuits between the prefrontal cortex and the amygdala become bigger. This is important because the stronger the link is, the less reactive you become. The relationship between these circuits and emotional reactivity is so strong that their size actually *predicts* the degree of a person's reactivity. Even when not faced with an emotional trigger, the amygdalas of adepts are 50% quieter.

In people who are very stressed, the opposite effect occurs. The amygdala grows in size. The connections between the regions of the prefrontal cortex that control emotions and the amygdala decay, allowing stress signals to proliferate.

What's particularly fascinating to me about this control circuit is that *information flows both ways*. In stressed people, the circuit sends signals from the amygdala to the prefrontal cortex, hijacking the brain's decision-making centers and paralyzing executive function.

But adepts hijack the hijacker. They use this same control circuit *in reverse*. The prefrontal cortex sends signals the other way, curbing the amygdala and shutting down the stress response.

The amygdala can also be regulated by the basal ganglia, especially the striatum. This engages with areas of the PFC to extinguish the amygdala's conditioned responses to fear. Once these structures deep in the brain are activated, emotion control becomes easier.

The One Percenters have recognized the necessity of emotional regulation for centuries. St. Abba Dorotheus, a 6th-century Palestinian abbot, advised his followers that, "Everything you do, be it great or small, is but one eighth of the problem; whereas to *keep one's state undisturbed* . . . is the other seven eighths." The Dalai Lama says, "The true mark of a meditator is that he has disciplined his mind by freeing it from negative emotions."

The Emotion Regulation Network therefore both *grows* those brain regions useful in regulating emotion and *shrinks* those that are not. This is the foundation of both a successful meditation practice and a happy life.

THE ATTENTION NETWORK

I place the Attention Network next because if we've learned to control our unruly emotions, we're then able to pay attention. When we're no longer distracted by emotion, our attentional focus is strong. This starts to grow the brain regions involved in the Attention Network. Among them are:

- Insula
- Somatomotor areas
- Prefrontal cortex centerline regions
- Posterior cingulate cortex
- Precuneus
- Anterior and mid cingulate and orbitofrontal cortex
- Anterior cingulate cortex

- Angular gyrus
- Pons
- Corpus Callosum

The somatomotor areas process pain, touch, and orientation of the body in space. The centerline of the prefrontal cortex is engaged by paying close attention. It can notice and correct mind wandering. The precuneus is involved in both self-regulation and happiness.

The other brain regions that make up the Attention Network are involved in regulating emotion and selfing, and directing the focus of attention. So the network as a whole is able to detect and correct Monkey Mind, control sensory distractions, keep attention focused, and suppress selfing.

An Expanding Brain in a Finite Skull

How can the brain increase in size if it's enclosed in a skull of fixed dimensions? This happens in an intriguing way. The neocortex has hundreds of folds. As the brain grows, the number of folds increases. Think of taking a bedsheet and wrinkling it. The more it wrinkles, the greater the amount of surface area you can fit into a smaller space.

The 10-dollar word that neuroscientists use for this folding is "gyrification." Gyrification allows for the addition of brain tissue within a fixed-volume skull. Research shows that meditators have more gyrification, resulting in faster information processing.

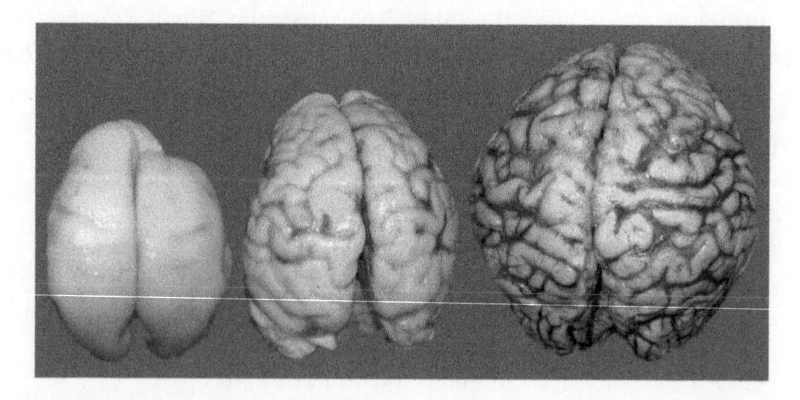

6.9. Midway through the development of the human fetus, the brain has no folds (left). At birth, it has developed substantial folding (center). Folding can continue throughout adult life (right).

Flexing My Brain

by Achmed Habibi

I began meditating for real 3 months ago. I get up at 4 A.M. and meditate till 6:30 A.M. Before that I used to meditate for around 20 minutes a few times a week, but I wanted to see if I could reach the same profound states I've heard people like Joe Dispenza and Tony Robbins achieve.

About a month back, I began to feel pressure in my head. It's about two inches above the bridge of my nose. It doesn't feel bad, but it's strong. Some days more so than others. I went to see my GP for my yearly medical check, but he couldn't find anything wrong with me; in fact, he said I was in great shape.

I read Dawson's stories in *Mind to Matter* and thought he might have insight. So I went to his EcoMeditation class and sat in the front row. That way I could ask a question.

Dawson asked me lots of questions in return. Then he told me that the pressure might mean that I am growing that part of my brain. I feel really good, so I suppose it's possible. I know that after a life of stress as a financial trader in London I am ready to take another, different journey with my life.

Others who've begun practicing EcoMeditation have told me similar stories. After a few months, they began to feel pressure inside their skulls. When I ask them where, they often point to similar areas.

These are places close to the surface, such as the temporoparietal junction and dorsolateral prefrontal cortex, in which meditation produces increases in brain tissue. As it grows, pushing against the skull to produce increased folding, you, too, may feel this pressure.

THE SELFING CONTROL NETWORK

In Chapter 2, we learned about "selfing," the tendency of the DMN to focus on the self, with its bias to suffering and negativity. The brain also has a Selfing Control Network that suppresses this activity.

Once we've learned to tame our emotions and focus our attention, we abandon our petty self-absorption and surrender into ecstatic states. When the Selfing Control Network is engaged, our obsession with selfing diminishes as we are drawn into unity consciousness with nonlocal mind.

In a classic study, Judson Brewer, one of the most innovative researchers in neuroscience, found a stronger connection in meditators between the dorsolateral PFC of Chapter 3 and the PCC. This neural wiring quiets the PCC, one of the two centers of the DMN. The Selfing Control Network enlarges a number of brain regions. These include:

- Ventromedial prefrontal cortex
- Inferior frontal sulcus
- Inferior frontal junction
- Anterior and mid cingulate
- Orbitofrontal cortex

After just 3 days of mindfulness practice, researchers found increased connectivity between the dorsolateral PFC and the PCC. Practitioners were building increased capacity for self-regulation just 72 hours after they began the practice.

But in highly experienced meditators with thousands of hours of practice, the connections were stronger. They could lose themselves in Bliss Brain more quickly and effectively. Their DMNs were also less activated than those of the novices. This results in the DMN being better controlled even during non-meditating states.

The medial PFC is involved with self-referencing behavior and quiets down in experienced meditators. The anterior cingulate helps manage impulses and emotions. Together with parts of the basal ganglia, it controls selfing by providing a future perspective rather than endorsing immediate behavior.

Enhanced connectivity in these brain regions enables the meditator to silence the incessant chatter of the selfing network, calm the DMN, and gain the mental space required for inner peace. This diminishes the Monkey Mind that pulls the novice out of meditative states.

THE EMPATHY NETWORK

Once these first three networks are developed, we're free of the self-concerned chatter that so distracts the novice meditator. The borders of our tent enlarge to embrace more than our skin-encapsulated ego. We feel one with the universe and one with everything in the universe.

This engages the Empathy Network. When we feel empathy, the same neural circuits light up as when we take care of children, friends, or other members of our tribe. These circuits grow stronger quickly when we regularly invoke compassion. And with the borders of our tent expanded to include the whole universe, empathy extends to infinity.

The Empathy Network also includes brain structures that perceive body signals such as pain and are crucial to emotional intelligence. The meditator feels the pain of another as though it is his own.

Once the Empathy Network lights up, it awakens the circuits associated with happiness and joy. The greater the connectivity between these brain regions, the happier and more altruistic meditators become. Among the brain regions important to the Empathy Network are:

- Insula
- Temporoparietal junction
- Anterior cingulate cortex
- Premotor cortex
- Nucleus accumbens

In compassionate adepts, the brain's insula begins to enlarge. The insula makes us aware of our internal emotional states and raises our level of attention to their signals. It also has rich connections to the heart and other visceral organs, allowing it to track and integrate signals coming from the body. In empathetic people, the insula responds strongly to the distress of others, just as though we were suffering ourselves.

Activation of the temporoparietal junction (TPJ) indicates that we can see things from the perspective of another person. This allows us to put ourselves in their shoes and take their needs into consideration.

6.10. Empathy is a neurological event, not simply an emotional state.

There's a part of the anterior cingulate that lights up only when we're contemplating actions that help others. It isn't activated by outcomes that favor only us. This region is also associated with impulse control and decision-making; we can choose win-win options rather than the desire-driven cravings of the nucleus accumbens.

When adepts are confronted by the suffering of others, the premotor cortex lights up. This means that the brain hasn't just noticed the distress of a fellow being; it's getting ready to take action.

In experienced meditators, the nucleus accumbens shrinks. This structure, which we looked at in Chapter 3, is active in desire and addiction. Deactivation of the nucleus accumbens through empathetic connection equates to a weakening of self-centered attachment.

Calming our emotions and focusing our attention, we're no longer driven by our wants and compulsions, and the brain circuitry associated with this part of the reward circuit begins to wither.

When training people in EcoMeditation retreats, I focus on the Empathy Network only after the first three networks are active. First I have them focus on self, then on just one other person. Only after that do we expand our compassion to the universal scale.

That's because thinking about other people can easily take us into mind wandering. People I love, people I don't, and the things that happened to cause those feelings. Trying to be compassionate toward people who harmed us can lead us out of Bliss Brain. So I activate the Empathy Network only after the Attention Network is engaged.

BRAIN CHANGE IS LIFE CHANGE

The fact that blissful states, practiced consistently, become blissful traits is a profound gift to us human beings. It means that we aren't condemned to live in the Caveman Brain with which evolution endowed us.

That practice can evolve our brains, some parts slowly, some parts quickly, is a remarkable innovation. In Chapter 8, we'll see how profoundly this is affecting the world.

The 22.8% growth in Graham Phillips's dentate gyrus in just 8 weeks is a clarion call to action. When you can develop your emotional regulation hardware—the foundation of happiness and inner peace—that fast, there's no reason to hesitate.

You don't have to accept the happiness level you know today. You can jump to a much higher level—with practice, all the way to bliss. And it need not take long. The Kirtan Kriya study found that just 8 weeks, 12 minutes a day, was enough to kick-start brain change.

We may look around us and be dissatisfied with parts of our lives—money, love, family, career, spirituality, weight, or health. Our ability to change them might be limited.

But that doesn't apply to our ability to change our brains. And when you change the brain that's living your life, you introduce a fresh perspective to the material circumstances around you. That's when the world around you begins to change too. Brain change sparks life change.

DEEPENING PRACTICES

Here are practices you can do this week to integrate the information in this chapter into your life:

- **Healthy Brain Journaling 1:** Marian Diamond listed the five essentials of a healthy brain: diet, exercise, challenge, newness, and love.

Write these five headings down in your journal at the start of the week. Each day, write down at least one action item you will take, in one area, to support your brain health. Here are two examples:

- *Newness.* This week I will play my French language audios in my car on the way to work and practice my vocabulary.
- *Exercise.* This week, when I go to the grocery store, I'll get off at the metro stop one extra exit away and walk the last half mile.

- **State Progression Exercise:** Use tactics for counteracting the brain's negativity bias and taking in the good. One example: Think of a person who annoys you. Now write down a list of their good qualities. As your mind is drawn back to their annoying ones, deliberately look at your list. Fill your mind with those good qualities as you make new neural links between that person and the list.
  You can do a state progression exercise on other stimuli that put you into an unresourceful state.
- **Healthy Brain Journaling 2:** Usually, "the hostile takeover of consciousness by emotion" is triggered by the same stimuli. In your journal, write down a list of 10 things that seriously annoy you about other people. Make them personal. You can make them as petty, vicious, and nasty as you like. Now fold that page over and date it a year from now. Don't look at it again till then. When you next review your folded page, a year or more later, notice how you've changed, without even trying to do so.
- **Noticing Head Pressure:** Notice if you've begun to feel pressure inside your skull when meditating. Write down exactly where you feel it. Now look at a brain image online. See if you can identify the brain region in which you're perceiving the pressure and read again about the function of that region.

EXTENDED PLAY RESOURCES

The Extended Play version of this chapter includes:

- Dawson Guided EcoMeditation: Cultivating Attention
- Andrew Newberg TEDx Talk on Finding Our Enlightened State
- Judson Brewer Video on the Default Mode Network
- Sara Lazar TEDx Talk on How Meditation Changes Brain
- Dawson Video Demonstration of Energy Changing Matter

Get the extended play resources at: BlissBrainBook.com/6.

CHAPTER 7

THE RESILIENT BRAIN

We've all heard of posttraumatic stress disorder or PTSD. Much of my research has been focused on the treatment of PTSD. With colleagues, I've conducted several randomized controlled trials of PTSD in veterans. We've shown that with using EFT, over 85% of them recover from symptoms such as flashbacks, nightmares, insomnia, and hypervigilance. Many also learn to meditate.

With a group of therapists and life coaches, I started the Veterans Stress Project in 2007. Over the years, we've worked with more than 20,000 veterans and their family members. It has been wonderful to see the changes in their lives. Here's a story from *Mind to Matter* of one veteran who learned EcoMeditation.

From the Battlefields of Iraq to Inner Peace

by Bryce Rogow

A lot of my friends say I'm a walking contradiction. On one hand, I'm a spiritual seeker; I studied meditation at a Zen monastery in Japan, I am a yoga teacher, and I've been learning mind-body medicine from some of the top healers around the world.

On the other hand, I'm a veteran of four combat deployments to Iraq as a corpsman, or medic, with the US Marines. After getting out of the military. I was diagnosed with PTSD, and after some time of feeling lost and hopeless, I embarked on a journey of self-discovery and healing, intent on learning the most effective techniques for cooling the fires of mental and physical distress.

In this letter I want to tell you about one of the most effective methods of healing mind and body that I've come across, a technique that's become an indispensable part of my daily routine and one of my favorite things to share with anyone suffering who wants to feel better now: Dawson Church's EcoMeditation, a practice that only takes a few minutes, yet can lead one into an elevated state of well-being commonly reserved for only the most advanced spiritual practitioners.

It's an ideal technique for any veteran suffering from PTSD—completely free, yet more effective in my experience than any of the pills or therapeutic techniques currently in use by the Veterans' Administration.

I'd like to tell you a little bit about my story, then describe in more detail the methods and benefits of EcoMeditation; it's my hope that anyone reading this will want to learn it for him- or herself and to include it in the list of therapies offered to veterans by the VA.

I graduated from Princeton and enlisted as a corpsman in 2003, the year we invaded Iraq and started a war that has come to define my generation—at least for those of us who served. I wanted to serve my country and also have a hand in liberating the people of Iraq from the tyranny of Saddam Hussein.

My first deployment with a US Marine Recon Battalion (the Marines' version of Special Forces) led me to the Second Battle of Fallujah in November 2004, a massive assault on a city that has been described as the most intense urban combat US forces have seen since Hue City in Vietnam.

All of us who deployed carry with us images that will stay with us for the rest of our lives, images we have to learn how to live with.

For me, the first image of that kind came after my first buddy in my unit was killed while digging up an IED, or Improvised Explosive Device—one of the homemade bombs insurgents bury in fields and roadsides.

After this friend was blown up (and several more were killed and wounded), his body parts were scattered in small chunks over an area the size of a football field. When I came on the scene, other members of his platoon were picking up his body parts, while the platoon corpsman treated the wounded. I was surprised when two of his teammates came to my Humvee asking for gloves and trash bags; the reason, it became clear, was that there was no "body" for a body bag, only chunks of flesh, bone, and organs that they were putting in 50-gallon black trash bags.

Of course, I was honored to sift respectfully through my friend's remains, but the images seared into my brain that night will stay with me every day for the rest of my life, and I began the process of divorcing myself from my own life and body, as a protective mechanism, because I too expected to end up as a trash bag full of chunks that would be sent home to my family, and yet I still had to function through another two and a half years of combat deployments.

My method for preserving my own mental function, in addition to becoming addicted to the painkillers we medics had available, was to accept the fact that I was already dead, and so I would constantly remind myself that nothing that happened to me would matter, because I was already dead.

PTSD has been described as an actual, physical injury to the brain, and I realize now that although we medics had the best first-aid training in world history, we had no "psychological first aid" to treat these mental wounds. And though we wore heavy body armor to protect us from snipers and enemy sharpshooters as we patrolled dusty, trash-strewn Iraqi streets, we had no "life armor" to protect us from the psychological toll of constant fear of death. No knowledge of the mind-body techniques to cool our nerves and access our inner healing resources when we needed them most.

When I received my honorable discharge from the US military in 2008, I was surprised to have survived the war. I expected a huge flood of relief when I was released from any possibility of future deployments, but that relief never came. I walked and drove around US cities with the same tense fear I'd experienced in Iraq.

I think that all that time I'd spent accepting the fact that I was already dead made me sort of a walking zombie among the living back home. Every person I looked at I would see as horribly disfigured, shot, maimed, bleeding, and needing my help. In some ways it was worse than being in Iraq, because the feelings were not appropriate to the situation and because I no longer had my buddies around to support me emotionally.

I spent a good deal of time heavily dependent on alcohol and drugs, including drugs such as Clonazepam prescribed by well-meaning psychiatrists at the VA, drugs that were extremely addictive and led to a lot of risky behavior.

However, I still had a dream of learning how to meditate and entering the spiritual path, a dream that began in college when I was exposed to teachings of Buddhism and yoga, and I realized these were more stable paths to well-being and elevated mood than the short-term effects of drugs. I decided that I wanted to learn meditation from an authentic Asian master, so I went to Japan to train at a traditional Zen monastery, called Sogen-ji, in the city of Okayama.

Many people think that being at a Zen monastery must be a peaceful, blissful experience. Yet though I did have many beautiful experiences, the training was somewhat brutal. We meditated for long hours in freezing-cold rooms open to the snowy air of the Japanese winter and were not allowed to wear hats, scarves, socks, or gloves. A senior monk would constantly patrol the meditation hall with a stick, called the *keisaku*, or "compassion stick," which was struck over the shoulders of anyone caught slouching or closing their eyes.

Zen training would definitely violate the Geneva Conventions. And these were not guided meditations of the sort one finds in the West; I was simply told to sit and watch my breath, and those were the only meditation instructions I ever received.

I remember on the third day at the monastery, I really thought my mind was about to snap due to the pain in my legs and the voice in my head that grew incredibly loud and distracting as I tried to meditate. I went to the senior monk and said, "Please, tell me what to do with my mind so I don't go insane," and he simply looked at me, said, "No talking," and shuffled off.

Left to my own devices, I was somehow able to find the will to carry on, and after days, weeks, and months of meditation, I indeed had an experience of such profound happiness and expanded awareness that it gave me the faith that meditation was, as a path to enlightenment, everything I had hoped for, everything I had been promised by the books and scriptures.

I am profoundly grateful to Shodo Harada-Roshi, a true modern Zen master, for facilitating that experience. However, after leaving the monastery, I realized that I would not be able to maintain that level of meditation on my own, that I would need faster and easier ways, and a better understanding of mind and body to make meditative practices a useful part of my life.

One day while researching on the Internet, I came across EcoMeditation. I simply read from the webpage and followed the steps, and within 2 minutes found myself activating all these healing resources and entering a state of profound relaxation and well-being that I'd previously achieved only after hours, if not days and weeks, of meditation.

I immediately signed up for a workshop with Dawson (which, to my incredible good fortune, was co-taught by his colleague Dr. Joe Dispenza). I was so excited for Dawson to lead us in the EcoMeditation, and when he did, I was not disappointed; instead I was convinced that his method was something the world, and veterans in particular, could use to experience profound inner healing and wellness without spending months at a monastery, which most of us do not have the freedom to do.

Although the media keeps us well-informed of PTSD, fewer are aware of the phenomenon of posttraumatic growth: the idea, and fact, that by making a conscious choice of not being a victim and of embarking on a path of healing, anyone who has experienced a trauma can come to find it as a meaningful event that leads them into a greater fullness of life. It's been said that suffering is the greatest grace, because it leads us to the pursuit of healing, which can take us to places beyond what we'd experience in our normal states of being.

But we need to know the tools of healing; those in distress need specific techniques and methods.

The VA and the US government have spent billions of dollars researching the best ways to treat the psychological wounds of our wounded warriors. Yet many of these techniques are themselves expensive, involving sophisticated technology and advanced drugs, and can be administered only by those with an MD or PhD after their name.

Although you may not have experienced the intense horror of close urban combat, in another sense we all do experience a milder form of combat exposure every day, as we are all bombarded by stressors in our modern society that activate our fight-or-flight response, evolved to save our lives in the event of grievous bodily injury, yet totally inappropriate when triggered by the many chronic minor stressors we face, from running late in traffic, to a hectic family life, to just reading the newspaper every day and being exposed to countless tragedies in third person.

Dawson Church's EcoMeditation and the associated Emotional Freedom Techniques (EFT) he teaches (based on Chinese energy medicine) are part of a growing movement away from drug regimens and toward "patient-administered, self-directed" therapy, techniques that can be taught to a veteran in half an hour at essentially no cost, techniques that the veteran can then draw on for the rest of his or her life, as a practice of daily mind-body hygiene, and something that offers rapid relief in periods of acute distress. I heartily recommend its immediate inclusion in the VA therapeutic armamentarium.

Bryce's story illustrates a phenomenon that is much less widely known than PTSD: posttraumatic growth. Many people who suffer shattering experiences are scarred for life, with little hope of recovery. But for others, shattering experiences prompt them to face their fears, transcend the horrors of the past, and become resilient. PTSD is not a life sentence.

POSTTRAUMATIC GROWTH

While PTSD grabs the headlines, news stories about posttraumatic growth are rare. Up to two thirds of those who experience traumatic events do not develop PTSD. This estimate is based on studies of the mental health of people who have undergone similar experiences. Studies of US veterans who served in Iraq and Afghanistan show this two-thirds to one-third split.

What's the difference between the two groups? Research reveals a correlation between negative childhood events and the development of adult PTSD. Yet some people emerge from miserable childhoods stronger and more resilient than their peers.

Adversity can sometimes make us even stronger than we might have been had we not suffered it. Research shows that people who experience a traumatic event but are then able to process and integrate the experience are more resilient than those who don't experience such an event. Such people are even better prepared for future adversity.

When you're exposed to a stressor and successfully regulate your brain's fight-or-flight response, you increase the neural connections associated with handling trauma, as we saw in Chapter 6. Neural plasticity works in your favor. You increase the size of the signaling pathways in your nervous system that handle recovery from stress. These larger and improved signaling pathways equip you to handle future stress better, making you more resilient in the face of life's upsets and problems.

7.1. Veterans at a retreat.

With some inspiring colleagues, I performed a study that touched my heart deeply. We examined the PTSD symptoms of 218 veterans and their spouses attending a weeklong retreat. The workshop included 4 days of EFT and other energy psychology techniques.

When they began, 83% of the veterans and 29% of the spouses tested positive for PTSD. On retesting 6 weeks later, these numbers had dropped dramatically: 28% of the veterans and just 4% of the spouses. Jenny, a veteran spouse who was on the verge of divorcing her husband before the retreat, commented tearfully, "After all these years, I finally got my Robby back from Vietnam."

While these are remarkable numbers, behind every one of them is a Bryce or Robby who has come back to their family and their life. The balance is tipped away from PTSD and toward posttraumatic growth.

This high cure rate is a gateway to a new set of possibilities for these veterans and their communities. Human potential that has been circumscribed by suffering is unleashed.

Saying Yes to Life in Spite of Everything: Viktor Frankl

The story of Viktor Frankl (1905–1997), an Austrian psychiatrist and neurologist imprisoned in concentration camps during the Nazi Holocaust of WWII, inspired the world after the war. By 1997, when Frankl died of heart failure, his book *Man's Search for Meaning,* which related his experiences in the death camps and the conclusions he drew from them, had sold more than 10 million copies in 24 languages. The book's original title (translated from the German) reveals Frankl's amazing outlook on life: *Saying Yes to Life in Spite of Everything: A Psychologist Experiences the Concentration Camp.*

In 1942, Frankl and his wife and parents were sent to the Nazi Theresienstadt ghetto in Czechoslovakia, which was one of the show camps used to deceive Red Cross inspectors as to the true purpose and conditions of the concentration camps. In October 1944, Frankl and his wife were moved to Auschwitz, where an estimated 1.1 million people would meet their deaths. Later that month, he was transported to one of the Kaufering labor camps (subcamps of Dachau), and then, after contracting typhoid, to the Türkheim camp where he remained until American troops liberated the camp on April 27, 1945. Frankl and his sister, Stella, were the only ones in his immediate family to survive the Holocaust.

In *Man's Search for Meaning,* Frankl observed that a sense of meaning is what makes the difference in being able to survive painful and even horrific experiences. He wrote, "We who lived in concentration camps can remember the men who walked through the huts comforting others, giving away their last piece of bread. They may have been few in number, but they offer sufficient proof that everything can be taken from a man but one thing: the last of human freedoms—to choose one's own attitude in any given set of circumstances—to choose one's own way."

Frankl maintained that while we cannot avoid suffering in life, we can choose the way we deal with it. We can find meaning in our suffering and proceed with our lives with our purpose renewed. As he states it, "When we are no longer able to change a situation, we are challenged to change ourselves."

In this beautiful elaboration, Frankl wrote, "Between a stimulus and a response there is a space. In that space is our power to choose our response. In our response lies our growth and our freedom. The last of human freedoms is to choose one's attitude in any given set of circumstances."

7.2. In recent years, record numbers have visited Auschwitz. The ironic sign above the front gate means "Work sets you free."

TRAUMA IS EVERYWHERE

It's not just veterans, crime victims, abused children, and accident survivors who come face-to-face with trauma. About 75% of Americans will experience a traumatic event at some point in their lives. Women are more likely to be victims of domestic violence than they are to get breast cancer. In any given year, 1 in 10 boys is molested and 1 in 5 girls, usually by a family member.

Sixty percent of teenagers witness or experience victimization in any given year. Close to half experience physical assault, and 25% witness domestic or community violence. During the 15 years of war in Iraq and Afghanistan, more Americans died at the hands of family members than were killed in the Middle East.

Look around you at the groups of people you encounter in everyday life. If you have 20 people on your work team, it's statistically likely that you're interacting with two or more incest survivors. Attend a men's support circle with 40 members and you're in a room with 30 people who've experienced a traumatizing event.

As we explored in chapters 2 through 6, the impact of these experiences is biological as well as psychological. In my book *The Genie in Your Genes,* I describe the role played by molecules called methyls, which attach to genes, inhibiting their expression. Certain molecular messengers are associated with depression, anxiety, and PTSD.

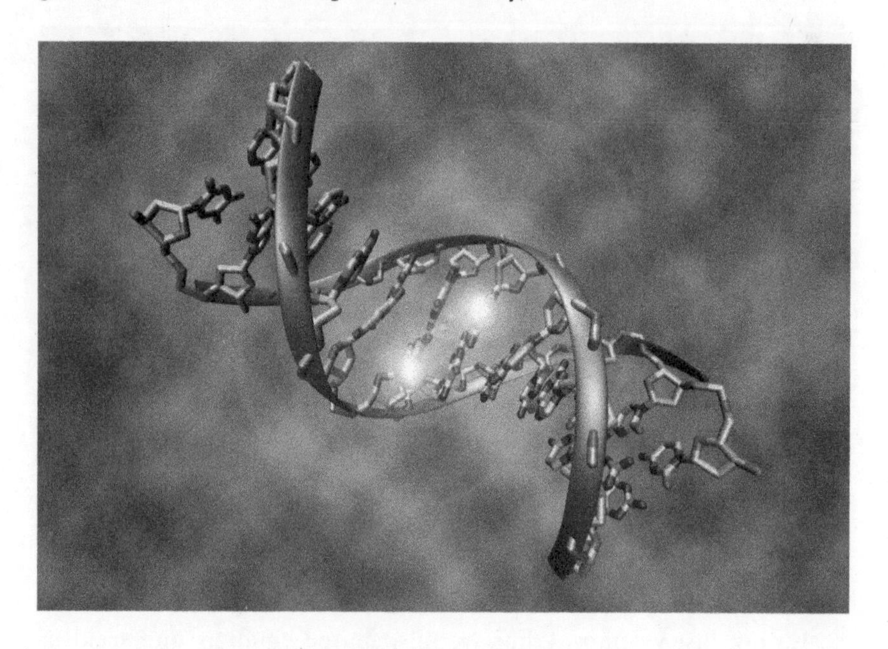

7.3. Methyl groups (bright spheres) adhering to a DNA strand.
See full-color version in center of book.

This is just one way in which childhood trauma results in "biological embedding" of negative experiences. Another is a class of tiny molecules called microRNAs. In a randomized controlled trial of veterans with PTSD, my colleagues and I found these bound to the DNA. The particular microRNAs we identified, called *let-7b* and *let-7c,* are linked to stress, and they inhibit the expression of beneficial genes.

Childhood trauma also inhibits brain formation. Adults who, as children, were deprived of nurturing have brains that are 8.6% smaller. The more severe the deprivation, the smaller the brain size. Epigenetically and neurologically, trauma shapes the matter of which our cells are constructed.

A TIME TO HEAL

In the Bible book Ecclesiastes (3:1–8), the prophet tells us, "To everything there is a season, and a time to every purpose under heaven." Among the "times" are "A time to kill, and a time to heal . . ."

For many of us, childhood was the time when much of our potential was killed under the onslaught of stress. Now that we're adults, it's time to claim our power and commit ourselves to the path of healing.

Adulthood is the life stage at which we can repair the wounds of the past. We face a choice between Door #1 and Door #2. With each challenging situation, we can act the role of the helpless child, walking through Door #1 to reenact mini-deaths of the past, or we can choose Door #2 and take on the mission of self-healing.

7.4. We have to choose daily between the two doors.

If we abandon ourselves to repeating the suffering of the past, we reinforce those neural pathways. That's why PTSD often gets worse over time; I call this the "dark side of neuroplasticity." Therapist Patt Lind-Kyle notes that the Default Mode Network of Chapter 2 is "strongly active in people with long-term trauma, childhood abuse, and other constrictive emotional patterns and experiences."

But if we choose Door #2, we practice the elevated emotional states we find in meditation, EFT, and other energy techniques. Neuroplasticity turns them into traits. We take the damage we suffered in the past and turn it into fuel for transformation.

As adults, we have power and we can make choices in a way we could not as children. The 5-year-old who is being molested can't run away from home and get a job waiting tables to pay for a safe apartment. An adult can. Adults may carry the impact of childhood trauma in their minds and brains, but they have power they lacked earlier.

Science is now showing us that we can heal those early-life deficits, not just in the software of our minds but in the hardware of our brains. Resilient adults use parts of their brains such as the ventromedial PFC and the dorsolateral PFC to control the amygdala. The more hours of practice they put in to control the amygdala, the stronger the effect. Practice can result in a 50% drop in amygdala activation.

Attentional control is the key. The father of American psychology, William James, called that which trains us to control our attention the most excellent education of all. We'll learn more about selective attention in Chapter 8.

Therapist Linda Graham, whose book *Resilience* contains 130 exercises to build a resilient brain, says, "Whether we're facing a series of small annoyances or an utter disaster, resilience is teachable, learnable, and recoverable."

Sages have told us this throughout the ages. Rabindranath Tagore, a Bengali poet born in the 19th century, asked, "Let me not pray to be sheltered from dangers, but to be fearless in facing them. Let me not beg for the stilling of my pain, but the heart to conquer it."

Helen Keller declared, "All the world is full of suffering. It is also full of overcoming." The Indian scripture Dhammapada (IX.122) says, "Drop by drop is the water pot filled. Likewise, the wise one, gathering it little by little, fills oneself with good." The 19th-century German philosopher Nietzsche proclaimed, "What does not kill me makes me stronger."

A research team at Northwestern University's science innovation center conducted a 15-year study of professional failure. They expected to find that it did lasting harm to participants' careers. Instead, to their surprise, they discovered that over the decade following failure, people "on average, perform much better in the long term" than those who enjoy early success, according to lead author Yang Wang.

George Bonanno, professor of clinical psychology at Columbia University, believes that resilience is actually the most common response to tragedy. Between 35% and 65% of "people who experience a disaster return to their normal routine shortly after the event and stay there."

This in turn produces positive effects on our biology. Bruce McEwen, head of the neuroendocrinology lab at Rockefeller University, states, "Healthy self-esteem and good impulse control and decision-making capability, all functions of a healthy brain architecture, are important" contributors to posttraumatic growth. He regards resilience as an internally generated state of being able to produce healthy epigenetic changes in one's own gene expression when under stress.

In Chapter 4, we saw the massive changes in brain activity that result from healing our past. In the microRNA study, after successful treatment, those stress molecules literally popped off the DNA of veterans. We found a statistically significant relationship between improvements in mental health and eight microRNAs.

7.5. The microRNAs study found correlations between these epigenetic molecules and stress symptoms.

The reality that we can produce these epigenetic changes to our chromosomes *using our consciousness* is one of the most exciting discoveries of recent years. In Chapter 8, we'll see the revolutionary implications of this superpower for humankind's future.

CONSCIOUSNESS AND LONGEVITY

While your genes, environment, and habits play a role in your health and longevity, the quality of your consciousness is paramount.

A large-scale study by Boston University School of Medicine followed 69,744 women and 1,429 men for decades: 30 years for the men and 10 for the women. It looked for "exceptional longevity," defined as living to age 85 or older. What made the difference?

Optimism!

The most optimistic people had a lifespan that was 11% to 15% longer. They had a 60% greater chance of reaching 85 when compared to the pessimists.

The results held true even after the study's authors adjusted their statistics to account for factors such as smoking, alcohol use, diet, education level, chronic disease, and exercise.

And that's just the *single* resilient trait of optimism. When you add the years that research demonstrates you gain from other qualities of consciousness, like love, joy, compassion, and altruism, the number goes up.

How high? My mentor, famed neurosurgeon Norm Shealy, did the math and concluded it was 40 years. That's a side effect of changing your consciousness through meditation, tapping, and other techniques for reducing stress. Studies show that meditators improve on a wide array of biomarkers including longevity. In one such study, wide-ranging epigenetic changes were found after just 20 minutes of meditation.

CHOOSING DOOR #2

I've always had physical challenges. My one leg is two inches shorter than the other, causing my hips to be misaligned and producing a big curve in my back. The medical term for that curve is "scoliosis." I've experienced back pain since my teenage years.

In my 30s, it was sometimes so severe that I'd be huddled against a wall in a fetal position. Any movement would produce such intense stabs of pain that I'd cry out like a wounded animal.

In my 40s, I learned how to manage my back condition using stretching, weight lifting, qigong, tapping, and yoga. The result is that I have since lived a relatively pain-free life. I've learned to thank my back when it gives me a twinge of pain, taking it as a signal from my precious body that I am working it too hard.

That was before the fire.

After the fire, we lived for a time in a funky French-themed hotel in downtown Petaluma. It had no elevator, only narrow stairs to give us access to our upstairs room.

One day while carrying a heavy box up the steps, I felt a sharp pain on the right side of my abdomen. I lay on the bed for a while and the pain went away. But it began to reappear regularly. Lying down or taking a nap usually relieved it.

Eventually, Christine insisted that I see my doctor, and he diagnosed it as a hernia. That's a tear in the muscles of the floor of the abdomen, resulting in the intestines poking through.

While the body has remarkable self-healing power, one thing it can't heal by itself is a hernia. Repair requires surgery.

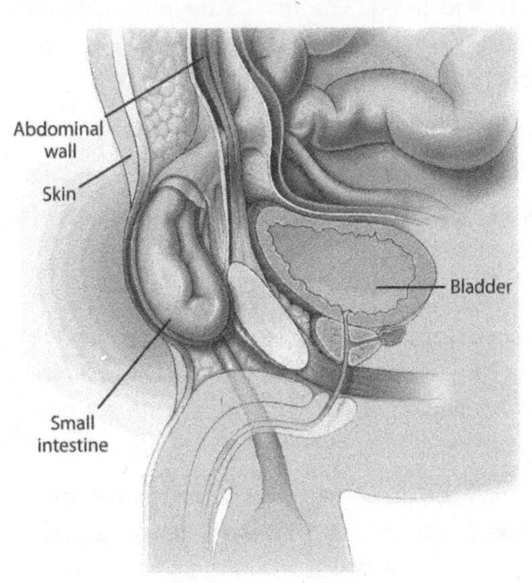

7.6. Inguinal hernia.

The hospital system was so overwhelmed treating injuries that had resulted from the fire that they couldn't schedule me anytime soon. I'd have to wait a couple of months.

I stayed as active as I could during the delay, bicycling, paddleboarding, and doing other exercises that didn't stress my abdomen. But every day, Christine and I had to cope with the realities of post-fire life.

On one of the worst days of my life, I found myself dragging heavy boxes across a parking lot to my borrowed truck. Most of our possessions were stored in a friend's office 30 miles away—and there were some I couldn't do without. As I made trip after trip from the office to the truck, my back and my abdomen competed for serving up the worst agony.

Yet I tapped on the way home, bolstered by that morning's meditation. Despite the hernia operation still being weeks away, I felt my mood going from despair to exhilaration. That's the power of resilience.

In the days, weeks, and months after the fire, I remained relentlessly, doggedly, determinedly optimistic. I gave thanks every day. I bathed my mind in Bliss Brain every morning. Choosing Door #2 makes you resilient—not just during a sudden catastrophe like the fire, but over the long haul of chronic pain.

A few months after the fire, we ran into an acquaintance at a party. She asked how Christine and I were doing, and we gave her the details of the challenges we had faced. The hernia operation. Money issues. Living like nomads, carting our meager possessions from one temporary shelter to another in plastic bins.

She was amazed at how cheerful we were. She recalled, "Yesterday I talked to my friend Lynne, another fire victim. She is still so angry and bitter."

That's where Door #1 takes you. We can let tragedy make us bitter, or we can celebrate our resilience.

In his book *Resilient,* psychologist Rick Hanson says, "Resilience is more than bouncing back from adversity. People who are resilient keep pursuing their goals in the face of challenges." Linda Graham advocates "coming to see ourselves as people who *can* be resilient, are *competent* at coping, and are competent about *learning* how to cope."

Our local Pulitzer Prize–winning newspaper marked the 1-year anniversary of the fire with a special issue. It contained heart-wrenching stories of people who were still living in trailers 1 year on. Of people sunk in despair. Of depression and PTSD. Of the slow rate of rebuilding.

The word "posttraumatic growth" did not appear even once in all those pages. The stories of people like me and Christine were not told. Reading the paper would give the reader no idea that recovery was even possible. Yet statistics show that posttraumatic growth may be the norm rather than the exception.

Christine and I visited our property several times in the months after the fire. Among the ashes we found fragments of beloved objects. Under the guidance of her art teacher, Carole Watanabe, Christine decided to make art out of the fragments. She arranged them around a mirror. Here's what she wrote as she prepared to hang it in her first art show.

7.7. The fire mirror. See full-color version in center of book.

The Fire Mirror

by Christine Church

The mosaic mirror called Fire Fragments tells a story of my past. This is an assemblage I made from pieces of pottery and jewelry. As I found each item in the burned rubble, I would wonder, "Out of thousands of things, why did this piece make it?"

Or "Look where we found this!" Objects that once had little or no meaning now became fellow survivors: symbols of strength and courage.

Top right: We never found the diamonds I was looking for, but we found a brown mass of jewelry melded together. In it was my mom's wedding ring! We threw piece after unrecognizable piece into a bucket. Our friend, Ray even brought his gold-panning equipment. We were so crazed that we didn't even use our masks or gloves.

Moving counterclockwise: My great-grandmother's Bavarian china, two Spanish Lladros from my dad's collection—"Girl with a Shoe" and "The Shoe" (top left and bottom left corners) with Bavarian china cup and saucer.

The cat inside the shoe represents our beloved Apple and Pierre. There is a Moroccan necklace that I wore in my 20s when I traveled to Spain and Morocco (bottom) and a duck made from the ashes of the Mt. St. Helens volcano.

Hope arises in the bottom right corner with a cup from a ceramic tea set my daughter Julia made in seventh grade, with a piece of melted glass inside. Glass melts at 2,500 degrees—that's how incredibly hot the fire burned!

The key to our house (right side) is in a symbolic pond where our nameless turtle found her name when we discovered she had survived the fire. Tubbs now lives out her days in a friend's backyard pond.

Rings and shards of pottery I had made are scattered throughout. A seed and a crystal from our new home are planted (lower right corner). Angels watching over us with love that night and always (top right and left corners).

I feel it was a blessing that these broken, burned pieces from the ashes could be repurposed into an art form. The process of putting it all together was a healing, and I felt a sense of completion and letting go when it was finished. It hangs in our house as a reminder of things lost but not forgotten. And turning disaster into beauty.

MONEY, STRESS, AND CHOICES

I remember the day I had to liquidate my youngest son's college account. American universities are expensive, and for 12 years I'd been making automatic payments out of my monthly paycheck to cover the expected bills.

But now, that pool of money was all that stood between me and financial insolvency.

The envelope with the statement from TD Ameritrade, the institution that managed the account, sat in the In tray on my desk for days, as I agonized about the decision. Do I or don't I? Is there another way? I picked up the phone, then put it down again. Picked it up, put it down.

We've all had moments when we had to make agonizing decisions. I made the call.

"Are you sure you want to withdraw these funds?" the account manager asked me.

"Yes," I replied, feeling about one inch tall.

I'd been at this point often over the past year. Christine and I had taken a $100,000 line of credit against our house in the spring, a few months before the fire. We had planned a remodel of the 60-year-old structure, with new floors, carpets, kitchen, and appliances.

Then came the first awful month when I couldn't make payroll. I had to transfer $10,000 from the home equity credit line into the business to keep my team and organization intact. While the company was doing profoundly transformative work in the world, it was in the middle of a transition from a professional training company to an online educational business, and income wasn't keeping up with expenses.

The next month was more awful yet, consuming another $20,000. Soon the line of credit was gone, but it still hadn't been enough.

Emotionally, I found it incredibly difficult to be working so hard yet losing money every month. It's like paying to work. Most of us very reasonably assume that if we put in our 40 hours a week, we'll get paid in exchange.

My situation was the opposite. I had to *pay in order to work.* Every hour I worked, cash was being sucked out of my bank account.

Then came the fire.

Our income vanished, while all our expenses marched on.

Supporters organized a GoFundMe campaign to help us recover. I phoned a friend who was an experienced fundraiser to discuss our financial plight. He asked me how much it would take to fill the hole. I told him: "At least $250,000."

There was a long silence.

Then he told me that if we asked for a quarter-million dollars, people would think it an impossible sum, feel hopeless, and not donate. He recommended that we ask for $50,000 instead. We did, and eventually we gratefully received $46,000 from our community.

Though the help was welcome, it wasn't enough. A short while later, I had to liquidate one of my two retirement funds. I put it into the business and breathed a sigh of relief, knowing that it would keep us funded for the next 6 or 7 months.

Within 60 days, it was gone. I had to clean out my other retirement account. We even had to empty the emergency fund the insurance company had advanced us for living expenses after the fire.

That still wasn't enough. That led to the truly awful day when I had to liquidate my son's college account. By the end of the year, we were over $300,000 in debt.

I focused on staying positive every day, despite the money issues, health challenges, and constant reminders of the fire. It took every bit of focus I possessed.

Six months after the fire, in the middle of the financial crisis, after one morning's meditation, I wrote these words in my journal:

> I woke up this morning feeling like I'm being cradled in the arms of God. The energy of Spirit fills every part of me with blessing. The universe radiates perfection all around me. I am cradled in this field of blessing. It holds us always in love and joy. It nudges us daily to experience the light and beauty at the core of our being.
>
> I realize that I'm 100% spiritually successful. I enjoy a life of attunement to the universe. Daily, I celebrate oneness between my human consciousness and the greater consciousness of which I am a part. That's the ultimate goal of every life, and I've lived it from the beginning.
>
> I choose to remind myself of this when I'm mesmerized by the things that haven't materialized in my material world after so many years of visioning and hard work.
>
> As I tune in to the universe's energy, I feel mine change in response. My thoughts become ordered and inspired. I start the day feeling optimistic, positive, enthusiastic, and creative.
>
> I embody prosperity. I attune daily to the energy of prosperity, as I have been doing for so many years. I know that material reality arranges itself around the signal that my consciousness produces.

The truth is that I am abundant in every possible way, including money. I choose to maintain the joy of that vibration. I celebrate every manifestation of success in my world, no matter how small.

I am grateful for my life just the way it is. I remain positive no matter what.

I have the most important thing attainable in any life: Oneness with the universe! I attune to its music every morning in meditation. My mind, cells, and energy field come into resonance with its song. I then move into my day inspired and aligned. What a wonderful life.

After writing those words, I decided to bask in the experience. I lay down in bed and visualized the experience turning from a delicious but intangible feeling into a hardwired neural fact in my body. For half an hour, I imagined synapses connecting around this new reality. I pictured making this inspired state the neurobiological trait of my brain.

In this way, throughout that year of financial challenges, I stayed relentlessly positive.

The business bounced back the following year, returning to profitability and posting its best annual income ever.

When the coronavirus crash occurred in 2020, life generously gave me another opportunity to stay positive. All the live workshops that paid our bills were canceled. We stared into a financial abyss.

One Sunday afternoon I was reading a study describing the search for a vaccine for MERS, the coronavirus that caused the 2012 epidemic. With a start, I looked at one of the molecules pictured in the research. Immunoglobulin, the coronavirus-munching antibody! My previous research had shown that these antibodies rose 27% after the weekend EcoMeditation workshop at Esalen, and soared 113% after a week of EFT.

With my EFT Universe team, I brainstormed the implications. We realized that we had to get this information to the world. Within 2 weeks, we had produced two free immunity meditations, a budget-priced program, and a live "virtual Esalen" workshop.

Members of our community were incredibly grateful for these immune-boosting resources. The free meditations reached over 100,000 people. Income from the paid programs kept the organization afloat.

LOCAL SELF AS HOST FOR NONLOCAL SELF

When you drop back into your daily life after meditation, you're changed. You've communed with nonlocal mind for an hour, experiencing the highest possible cadence of who you are. That High Self version of you rearranges neurons in your head to create a physical structure to anchor it. You now have a brain that accommodates both the local self and the nonlocal self.

My experience has been that the longer you spend in Bliss Brain, whether in or out of meditation, the greater the volume of neural tissue available to anchor that transcendent self in physical experience.

Once a critical mass of neurons has wired together, a tipping point occurs. You begin to flash spontaneously into Bliss Brain throughout your day. When you're idle for a while, like being stuck in traffic or standing in line at the grocery store, the most natural activity seems to be to go into Bliss Brain for a few moments.

This reminds you, in the middle of everyday life, that the nonlocal component of your Self exists. It also brings all the enhanced creativity, productivity, and problem-solving ability of Bliss Brain to bear on your daily tasks. You become a happy, creative, and effective person.

These enhanced capabilities render you much more able to cope with the challenges of life. They don't confer exceptional luck. When everyone's house burns down, yours does too. When the economy nosedives, it takes you with it.

But because you possess resilience, and a daily experience of your nonlocal self, you take it in stride. Even when external things vanish, you still have the neural network that Bliss Brain created. No one can take that away from you.

DEEPENING PRACTICES

Here are practices you can do this week to integrate the information in this chapter into your life:

- **Posttraumatic Growth Exercise 1:** In your journal, write down the names of the most resilient people you've known personally. They can be alive or dead. They're people who've gone through tragedy and come out intact. Make an appointment to spend time with at least two of the living ones in the coming month. Listen to their stories and allow inspiration to fill you.

- **Neural Reconsolidation Exercise:** This week, after a particularly deep meditation, savor the experience. Set a timer and lie down for 15 to 30 minutes. Visualize your synapses wiring together as you deliberately fire them by remembering the deliciousness of the meditation.
- **Choices Exercise:** Make 10 photocopies of illustration 7.4, the two doors. Next, analyze in what areas of your environment you often make negative choices. Maybe it's in online meetings with an annoying colleague at work. Maybe it's the food choices you make when you walk to the fridge. Maybe it's the movies you watch on your TV. Tape a copy of the two doors illustration to those objects, such as the monitor, fridge, or TV. This will help you remember, when you're under stress, that you have a choice.
- **Posttraumatic Growth Exercise 2:** Write down in your journal an event from the past that hurt you deeply. Now stretch your imagination by writing down a list of the ways in which this helped you. Little detail is required; just a few words are enough. For instance, under the heading of "Terrible Divorce," you might list "(a) made me conscious of the value of saving money, (b) I became better at spotting potential disasters, (c) I found the wonderful therapist who has supported me ever since, (d) I discovered who my true friends were," and so on.

EXTENDED PLAY RESOURCES

The Extended Play version of this chapter includes:

- Dawson Guided EcoMeditation: Building a Resilient Brain
- The Veterans Stress Project
- Audio Interview with Linda Graham, Author of *Resilience*
- Tap-Along Video Session of Dawson Working with a Childhood Abuse Survivor
- Free Chapters from Dawson's Book *EFT for PTSD*

Get the extended play resources at: BlissBrainBook.com/7.

CHAPTER 8

FLOURISHING IN COMPASSION

While the focus of this book is on the changes produced by meditation, in reality *everything* we do changes our brains. Classical violinists add volume and speed to the parts of the brain associated with manual dexterity. Elite athletes increase the size and signaling capacity of those brain regions that handle coordination and proprioception. Race car drivers tune the parts of their brains responsible for reaction time.

We're all changing our brains all the time. Everything we direct our attention to sends signals through neural pathways. Running our fingers through our children's hair. Savoring a sip of fine wine. Reacting to the news. Cheering our favorite sports team. Watching pornography. Gambling with money. Playing chess. Worrying about our retirement fund. Playing video games. Gossiping. Work. Hobbies. Getting stressed out about ________________________ (insert your pet peeve here).

Once this concept sinks deep into our bones, the next step is to take conscious control of the process. We want a brain remodeled for joy, not the misery that most brains default to. By *selecting where we direct our attention,* we can install that upgrade in the hardware of our brains.

FROM SURVIVAL TO BLISS

In ancient times, the most important function of our brains was to keep us alive. Caveman Brain was oriented to survival: Remembering the bad stuff that happened yesterday. Scanning the horizon for bad stuff that might happen tomorrow. Avoiding threats. Seeking opportunities. Remodeling the brain for reactivity, suspicion, paranoia, and fear gave us a survival edge.

The name of our species is *Homo sapiens.* "Sapiens" means "wise." For Caveman Brain, "wise" meant fighting enemies, dodging predators, chipping flints, digging for roots, finding water, making babies, and sparking fire. Wise meant the ability to lift a chunk of dried excrement to your nose and know what animal it came from. Wise meant being able to hunt and kill your dinner.

For 20th-century humans, being "wise" meant something different. Wise meant getting good grades, finding a job, getting married, raising kids, having fun, staying healthy, avoiding bad habits, and saving enough money for the golden years.

For the new human, "wise" means something very different. It means being self-aware. Aspiring to your full potential. Noticing your character flaws. Identifying what makes you happy and what makes you miserable, then taking action to nudge yourself toward the happy end of the continuum. Living as a spiritual being on a physical path. Taking care of all your material needs just like your ancestors—then going beyond those to Bliss Brain.

Bliss Brain is the brain of the future. It's beckoning us to the next stage of evolution, and a wisdom that transcends the best of our ancestry.

SELECTIVE ATTENTION

Selective attention is the process of *bringing desired experiences to the foreground of consciousness* while delegating others to the background. When we train ourselves to direct our attention *deliberately,* we are able to shift our emotions in a positive direction even amid the distractions and annoyances of everyday life.

The trick is to practice the subject-object shift we learned in Chapter 3. Even when you're being buffeted by a strong emotion like the fear that served Caveman Brain so well, selective attention allows you to change. You can shift to the perspective of a witness, downregulate the intensity of feeling, and move yourself to a positive state.

Selective attention doesn't mean denying the problems you face. It isn't avoidance; it's making a choice from among a palette of options. While acting on fear was the best option for Caveman Brain, a conscious person examines all the options before selecting a thought, feeling, or behavior. By using the power of selective attention, we reshape our brains.

Van Gogh and the Sunflowers

by Carole Watanabe

I was extremely ill with Lyme disease. I was in a coma part of the time, and paralyzed. I could not talk. Everyone around me thought I was dying. I thought I might be dying too.

Then I remembered a vision I had as a child of standing in a field of sunflowers and painting like Van Gogh. He was my idea of the ultimate artist. Out in nature, painting something glorious.

Eventually I recovered enough to use my right arm. I wrote on the wall next to my bed: "Paint sunflowers in France."

I thought, "There's a new turn in the road; it's going to lead to your healing."

I recovered enough to stand and paint for 30 minutes, after which I'd have to rest for a couple of hours. As soon as I could, I traveled to France.

I wound up buying a funky house in the village of Soreze, outside of Toulouse. It was the least expensive house in the village.

But it was surrounded by sunflowers. It was like a beautiful jewel that dropped into my lap from the whole bad experience of Lyme disease.

Standing in the fields, painting sunflowers, felt like a yellow infusion of light permeating my body. I fulfilled my childhood vision. I healed.

Carole, my wife Christine's art teacher, used selective attention to focus on her vision of sunflowers even when she was near death. Her vision took her above her mortality. Carole's sunflower story is one of many shared in my Master Manifestor's Club online community. She's a Master Manifestor. Many synchronicities have shaped her life and those of her students.

8.1. Sunflowers surrounding a European village.

YOUR EVERYDAY SUPERPOWER

The ability of human beings to change the anatomy of their brains is awe-inspiring. We direct our attention selectively, certain brain regions light up, others go dark, and the neural remodeling of Chapter 6 begins.

Most people use this superpower unconsciously. They live their lives without reflection. They ascribe the way they are to forces outside themselves, and repeat their conditioned behavior day in and day out. They believe that they're unhappy because of the boss, angry because of the corporations, frustrated because of the government, resentful because of the flaws of their loved ones, and miserable because of money. Unaware of their power to direct their attention, Caveman Brain shapes their neural network around their peeves.

Conscious people take control of the process. They use selective attention to *direct their thoughts and feelings*. They decide to fill their minds with positive thoughts, then take action on that goal. They use this superpower consciously. Their lives improve as their brains are reshaped.

Identical twins are one of the favorite subjects of researchers. That's because, at birth, they have identical genes. Their brains are identical, as are their bodies. It seems logical to assume that biologically, they will lead nearly identical lives, year after year and decade after decade.

Yet they rarely do. Their health and happiness levels drift apart over time. One might become healthy and the other unhealthy. Or one happy and the other unhappy. Or one stressed and the other relaxed. On average, *identical twins die more than 10 years apart.* That's the power of brain and body remodeling.

8.2. Identical twins look the same at birth.
See full-color version in center of book.

As individuals we can use our superpower consciously, to upgrade our brains and lives, or let that superpower run in the background, replicating the conditioning of our past.

If we become masters of selective attention, uncompromising in our commitment to love, joy, and transformation, what happens to our brains? If we decide to dial up our superpower to the max, *how big an upgrade* can it produce in our brains?

8.3. But they diverge over the years, and may look quite different at age 60. While chronologically the same age, biologically some twin pairs are 11 years apart. See full-color version in center of book.

HOW FAR CAN NEURAL REMODELING GO?

Chapter 3 showed how we activate parts of the brain during meditation, while shutting down the Default Mode Network (DMN) of Chapter 2. Chapter 6 showed how this grows the Enlightenment Circuit while shrinking the stress network. Just how far can this process of neural remodeling go?

If you decide to be like one of the X-Men superheroes and develop your superpower, how big can it get? Now that you know the power of consciousness to shape the anatomy of your brain, how far can you take the process? Does neural remodeling have limits, and if so, what are they?

The fMRI studies comparing groups of meditators give us a guide. In Chapter 6 we saw that even a few hours of Awakened Mind begins to produce changes in the brain. When people *sustain alpha consistently* in a daily meditation practice, their brain anatomy changes measurably in just 8 weeks.

In adepts, the changes are huge. Those who've meditated for 10,000 hours or more show complete attentional control, suppressed selfing, and shrinkage of their stress networks. Bliss Brain is their default.

But the brain keeps changing even beyond that point. Richard Davidson measured how easily the brain's "fire alarm," the amygdala, could be activated in meditation adepts. He compared a group of monks who had spent an average of 19,000 Total Lifetime Hours (TLH) in meditation with an even more experienced group. These super-meditators had 44,000 TLH under their saffron belts. That's the equivalent of 10 years of solid practice 12 hours per day. He found "a staggering 400% difference in the size of the amygdala response" between the two groups.

This means that the brain's stress networks *continue to contract throughout the lifetime* of the meditator.

Bliss Brains also age more slowly. The thickness of gray matter is a marker that allows scientists to measure the biological age (as opposed to the chronological age) of a person's brain. Mingyur Rinpoche, a 41-year-old adept studied by Davidson, had the brain age of a 33-year-old. Even after tens of thousands of hours of practice, the brain keeps changing.

8.4. Mingyur Rinpoche.

This research gives us the answer to the key question of how much change you can produce: If you exercise your superpower consistently, *there's no upper limit.* Your brain can keep on evolving throughout your entire life.

So why not max out your potential? Go for the gold. Create the happiest brain possible.

The Side Effects of Bliss Brain

Like every human being, I have good qualities and bad qualities. One of the latter is impatience. I've always wanted this flaw fixed—and right now!

Christine and I visit gorgeous resorts each year as we look for the most nurturing settings to offer participants at our Life Vision Retreats. We booked a weekend trip to San Diego to check out Rancho Bernardo resort, a site we were considering.

After enjoying the weekend, we arrived at the airport on Sunday evening for our return flight. It was delayed for 4 hours because of fog back in San Francisco. We waited patiently in the lounge.

Then the airline announced that the flight had been canceled. We joined a long queue of disgruntled passengers waiting to rebook on the next available flight.

We eventually got to the head of the line and talked to an attendant whose badge told us her name was Sally. She told us that all flights were full until Tuesday. Not only that, we would not receive compensation for a hotel or rental car because the cancellation was due to the weather and not the fault of the airline.

I turned to Christine and proposed, "How about another day's vacation in San Diego, then we rent a convertible, and drive back to San Francisco?" She lit up at the thought of the fun we would have on the road trip.

I smiled, thanked Sally, and prepared to turn away, but she remarked, "I can't believe how patient you are."

"Sally, I really appreciate how you're doing your best," I responded, "and I'm grateful to you for taking care of us." Her eyes became moist as Christine and I walked away, hand in hand.

Driving back home with the top down, enjoying the green hills of a California spring, I had lots of time to reflect. I realized that *nothing* on the trip had triggered my flaw of impatience. I turned to the ultimate reality tester: my wife. "Hey, Christine," I wondered, "have you noticed any time on this trip when I was impatient?"

She thought for a moment and responded, "I can't think of one."

"Can you remember the last time I was impatient?" I asked.

She had to go back a year or two to recall an incident.

I realized with a start that I am no longer a generally impatient person. Somehow impatience has slipped into the background of my experience. Annoyances that used to make me impatient don't trigger me now.

This is likely a side effect of spending lots of time in Bliss Brain. My happiness circuits are sufficiently developed, and my suffering circuits sufficiently silenced. The brain tissue devoted to suffering has shrunk as a proportion of my brain as a whole. The ratio between selfing neurons and happiness neurons has flipped in my favor. State change has led to trait change, and the trait of impatience is no longer me.

I find myself slipping spontaneously into Bliss Brain several times each day, just as though I'm meditating. That state is so close to the surface of my consciousness that it pops to the top now and then.

When I have a long drive ahead of me, I often go into Bliss Brain with my eyes open. I feel the pressure inside my head as my consciousness connects with nonlocal mind. I intensify the experience, and enjoy the opportunity to inhabit a mindful state for a while.

This is likely to be stimulating additional growth in the Enlightenment Circuit, taking me ever further from the Caveman Brain I inherited from my ancestors.

So leverage your superpower to the max. Engage that Enlightenment Circuit regularly and intensely. Your payoff? A *much* happier brain to live in.

While neural remodeling gives us happier individual lives, what happens when whole groups of people start doing it? What changes when consciousness begins to shift on a global scale, and how does this play out in social and historical events?

FLOURISHING CONSCIOUSNESS, FLOURISHING WORLD

Evolution by natural selection proceeds slowly. For instance, it took about 1 billion years for the earliest ocean-dwelling life-forms to evolve into land dwellers. It took another 250 million years before the first mammals appeared.

But there are exceptions, times when evolution takes swift and enormous leaps forward. Futurist Jean Houston calls these periods "Jump Time."

A famous jump called the Cambrian explosion occurred 535 million years ago. In a very brief period of time by evolutionary standards, every major body plan in the fossil record appeared. Before the Cambrian period, there were no starfish, lice, jellyfish, or clams on earth. Suddenly there were.

8.5. New life-forms that emerged during the Cambrian explosion. See full-color version in center of book.

Some jumps are biological while others occur in consciousness. When the consciousness of groups of people changes, tectonic shifts in religion, society, art, and culture can occur. Often this begins with just a few individuals, but when critical mass is reached, an entire civilization can change its collective mind quickly.

In 1890, the number of countries in which women were able to vote was 0. Then in 1893 the first country, New Zealand, gave women the vote. Within 40 years, after thousands of years of political suppression, women could vote in virtually every democracy. We as a human species changed our collective minds. Consciousness shifted, and the social wisdom of 50% of the human race came online.

There are many other examples. Only in relatively recent history did the human race change its collective mind about child labor.

In 1800, child labor was uncontroversial. In the coal mines of Britain, 8-year-olds pushed heavy tubs inside underground tunnels. Sweating and breathing black coal dust, they emerged from the pits black with grime. They died young.

Then we changed our minds, and in just a few decades, child labor was banned.

We changed our minds about slavery, and even though it had been around since the dawn of society, it disappeared in a few decades.

We're in the process of changing our minds about the legalization of drugs. About doctor-assisted suicide. About the death penalty. About gay marriage. About gun control. About racism. About universal health care.

Each of these is an example of mass human consciousness changing. Here I use the word *consciousness* in the sense of the collective picture of reality we hold in our heads as a species. In my book *Mind to Matter* I summarize over 400 studies showing that when consciousness (our internal reality) changes, the material world (external reality) changes along with it.

It's also noteworthy that each of these historical changes involved *people becoming more compassionate.* We began to care about the well-being of people other than ourselves; in these instances children, slaves, and disenfranchised women. People in power, with no incentive to relinquish it, were able to *mentally place themselves in the shoes of those less privileged than themselves.* They then voluntarily shared their power.

We're in the midst of a huge jump in both consciousness and material reality. By virtually every standard, global well-being has been rapidly improving. The improvements are summarized in a brilliant book called *Enlightenment Now: The Case for Reason, Science, Humanism, and Progress* by Steven Pinker. He shows how much of human society was static for centuries or millennia, but that a huge jump has occurred in the past 300 years.

Consider just a few of hundreds of examples:

- Between 2000 and 2015, 1 billion people (one seventh of the world's population) escaped poverty.
- The life span of the average global citizen has doubled in the past century.
- In developed countries, the proportion of government spending on social projects such as education and relief for the poor rose from under 2% at the start of the 20th century to an average of 22% today.
- Deforestation of the Amazon rainforest dropped by 80% in the past 2 decades.

- Wars between Great Powers, those able to project force beyond their borders, were continuous 450 years ago. They have declined since then and are nonexistent today.
- The world's wealth has tripled in the past 4 decades, and that's including the crashes of 2000, 2008, and 2020.
- From 1921 to today, US traffic accident fatalities have fallen 24-fold. Accidental deaths from all other causes have fallen as well.
- For the world as a whole, human rights have improved over the past 70 years.
- Google searches for homophobic, racist, and misogynistic terms have decreased by at least 80% since 2004.
- Despite the interruption of horrific wars, each century has been more peaceful than the one that preceded it.
- Since 1909, Intelligence Quotient (IQ) scores have been rising in all parts of the world by an average of 3 points per decade.
- From 1960 through today, the American poverty rate, measured by income, has fallen by more than half. Measured by what people consume, it has declined by 90%.
- World carbon emissions per dollar of GDP have dropped by over 50% in the past 50 years.
- Child labor, infant mortality, maternal mortality, and domestic violence have all been declining globally.
- The proportion of the world's people without access to clean air and water has roughly halved in a decade.

These improvements are only visible if you take a 30,000-foot overview. They can't be seen from the ground. When our minds are caught up in a huge global drama like the housing crash of 2008, it's virtually impossible to see that this is a little blip in a big upward arc.

Watching news reports about the millions of people infected with the coronavirus in 2020, it was equally difficult to remember that global health is on an upward trajectory. When the virus crashed the stock market, and US unemployment soared to 20% of the workforce, the upward slope on a 10-year graph seemed like a cruel joke.

Yet at such times of trial, it's more important than ever to disentangle your mind from fear, take a step back, and focus on the big picture. Here are some graphs depicting the 30,000-foot view.

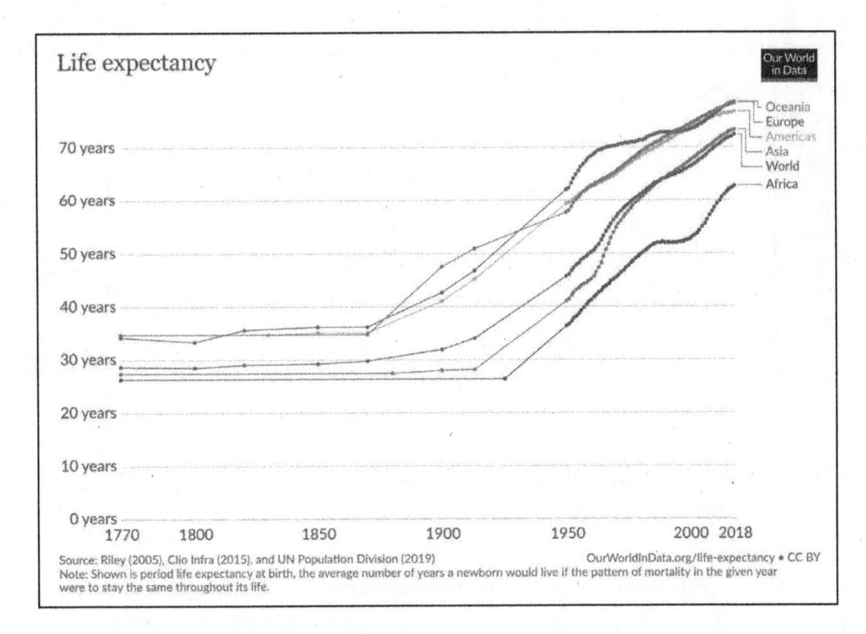

8.6. Life expectancy by world region from 1770 through 2018.

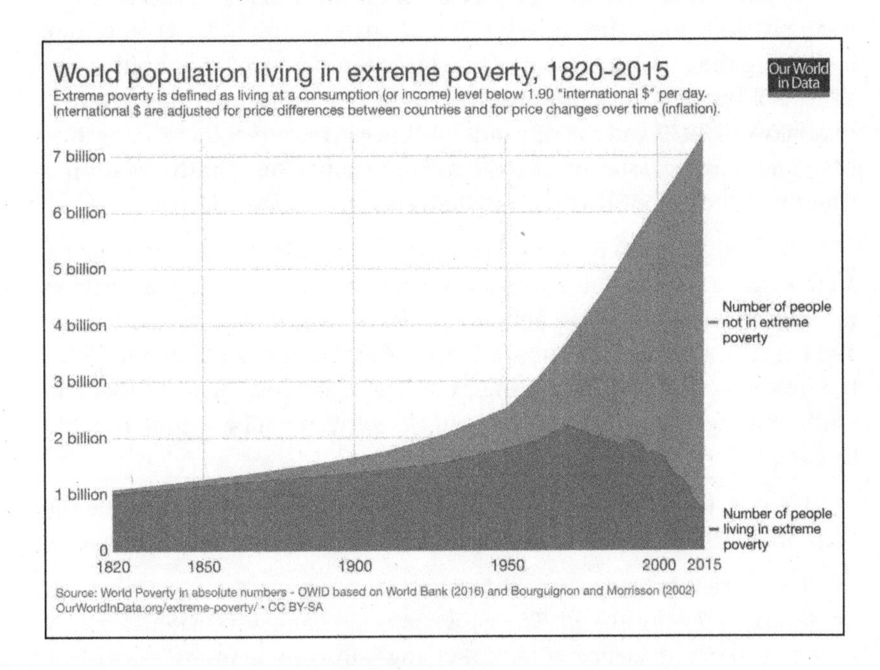

8.7. World poverty 1820 to 2015.

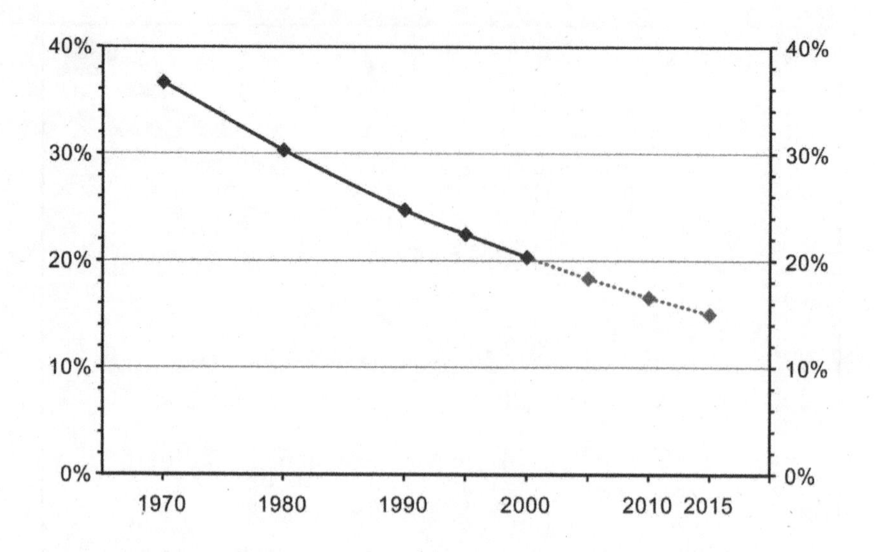

8.8. World illiteracy halved between 1970 and 2015.

Empirical analysis shows that we are in the midst of an explosion of spiritual, physical, and material well-being, though you can't detect this jump by reading the news. In fact, you'd be misled into missing it altogether. An archive of the frequency of words used by the media in 130 countries finds that between 1979 and today, words with negative connotations (like "horrific" and "terrible") steadily increased. At the same time, positive words like "improve" and "good" declined, as did the happiness level of US citizens.

The media depicts the opposite of what's actually happening, which is an upsurge in human flourishing. It coincides with massive global changes in consciousness, like the abolition of slavery, regulation of child labor, and the rise of democracy and women's rights. For the last 300 years or so, the human collective consciousness has been changing its mind about all kinds of things. Many of these changes were driven not by self-interest, but by compassion for others.

We now have the scientific tools to measure what happens inside the brain when consciousness changes. We know that it produces brain change.

Therefore it is likely that *the brains of human beings have been changing for the last 300 years as well.* The above radical changes in material reality may be external evidence of radical changes in consciousness, our collective internal mental reality, as well as parallel changes in brain anatomy.

HAS COMPASSION BEEN DRIVING BRAIN EVOLUTION?

Was there an uptick in meditation 300 years ago? Not that we can measure. But there was a steady increase in social reforms that give evidence of compassion for others. The research on the different types of meditation summarized in Chapter 4 shows that *compassion moves the needle like nothing else.*

It is therefore possible that compassion has been changing the brains of human beings little by little for centuries. We can't go back and conduct MRI scans on abolitionists or suffragettes, but the rapid shifts evident in social norms suggest that brain change on a large scale may have been occurring.

As societies have become much more compassionate in the past few decades, brain evolution might have picked up speed. Tantalizing evidence for rapid change found in metrics like Google searches for racist and homophobic terms—down 80% in less than 20 years—suggests this possibility.

If this hypothesis is true, social change will continue to accelerate, as compassionate brains create ever-more-compassionate brains. Human society could look completely different in a matter of years, not decades. We could be on the threshold of the most compassionate and enlightened civilization in history.

The idea that emotions like compassion might be driving evolutionary change is not new. Charles Darwin became interested in emotions later in his career. In his 1871 book *The Descent of Man,* he wrote: "sympathy will have been increased through natural selection; for those communities, which included the greatest number of the most sympathetic members, would flourish best . . ."

A large-scale scientific review identified several ways in which compassion has played a role in evolution. It suggests that our distant ancestors "likely preferred mating with more compassionate individuals—a process that over time would increase compassionate tendencies within the gene pool."

The review also points out that compassion confers an evolutionary advantage to a society because it "enables cooperative relations with nonkin." Compassionate societies flourish more. The process might have been nudging evolution along since prehistoric times.

We're not simply *in* Jump Time. With our compassionate new brains we may actually be *creating* Jump Time. In my next book and a series of scientific papers, I will be exploring the implications of this hypothesis for every facet of human endeavor, from education to medicine to law to science to business to technology to art.

THE BREAKNECK PACE OF EVOLUTIONARY CHANGE

The time frames for evolution and social change provide a context for how extraordinary your brain-remodeling superpower actually is. The Cambrian Explosion took over 12 million years from start to finish. That's quick in evolutionary terms.

The size of the human brain tripled in the 3 million years between the invention of the first stone age tools and today. That's lightning-fast in contrast.

It took 40 years to get women the vote. And here you are, you super-manifestor you, able to produce major change in your brain in just 8 weeks.

This is an event unparalleled in evolutionary history. Never before has a species been able to change the anatomy of its own brain with its consciousness. This superpower is unique to *Homo sapiens*, and it is changing the world.

So how many human beings are engaged in this grand experiment? What percentage of the population are Bliss Brainers, and how are they contributing to Jump Time?

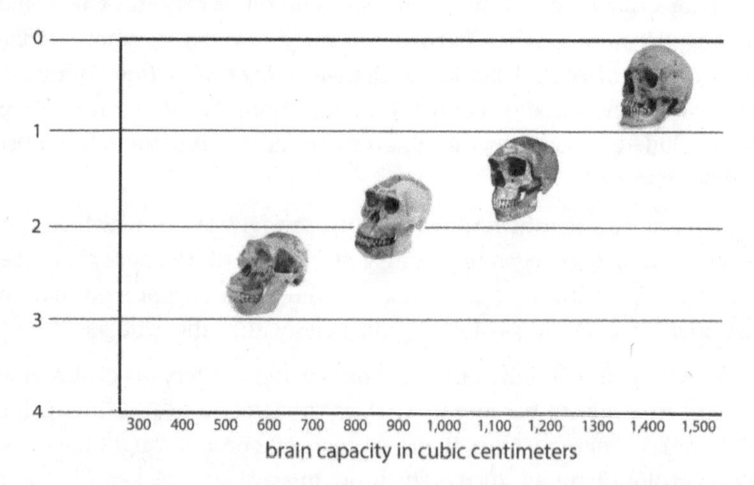

8.9. Brain capacity in cubic centimeters before and after humans began using tools, a span of 3 million years. Left: *Australopithecus africanus*. Center left: *Homo habilis*. Center right: *Homo erectus*. Right: *Homo sapiens*.

WHAT PERCENTAGE ARE MEDITATING TODAY?

In Chapter 4, we looked at evidence suggesting that throughout history, around 1% of people have actively sought enlightenment. In Tibet, the most spiritual region on earth, the number is 12%. What's the comparable number for Western societies today? And for the world as a whole?

One answer comes from a large-scale survey undertaken by the US National Center for Health Statistics. It found that the number of meditation practitioners tripled between 2012 and 2017, going from 4% to 14% of the population. That's a huge increase in a very short period. It continues to rise.

Not only adults but also children are increasingly practicing. Among 4- to 17-year-olds, the percentage rose from fewer than 1% to over 5%.

8.10. Mother and daughter meditating together.

According to the European Values Study, almost half the population prays or meditates at least once a week. Meditation has long been a cornerstone of religious practice in China, India, and other Eastern countries, which together make up half the world's population.

These trends are likely to accelerate in the decades ahead. People in Bliss Brain feel good, their brains awash in the pleasurable neurochemicals of Chapter 5. These drugs motivate them to get more. This makes them feel even better, so they expand their practice. Others notice the happiness of Bliss Brainers, and try meditation too. It spreads throughout society in a self-reinforcing virtuous cycle.

Yesterday, Bliss Brain was the exclusive domain of the One Percent. Today it's spread to the Fourteen Percent. Tomorrow, the majority. That's how we reach the tipping point of history.

AN INFINITE RESOURCE FOR A FINITE PLANET

We need this enhanced human potential to solve the problems humankind faces today. The resources of the planet are finite. In his second inaugural address in 1985, US President Ronald Reagan said, "There are no limits to growth and human progress." Yet there are.

Even though global wealth has been increasing exponentially, we live on a finite planet. We need both the consciousness and the economic systems that will allow our species to live in equilibrium with the rest of nature. Futurist David Houle calls this the "Finite Earth Economy."

Climate science tells us that we're facing a potential 3-foot (1-meter) rise in sea levels this century. In one low-lying country, Bangladesh, a 3-foot rise will put 20% of the country underwater, and 30 million people will become climate refugees.

The average elevation of the land in the US state of Florida is 6 feet (2 meters) above sea level. Sea level rises of 10 feet (3 meters) are predicted between 2050 and 2075, and 50 feet (15 meters) by 2500. Climate change is only one of the wicked problems that our evolving brains are required to face.

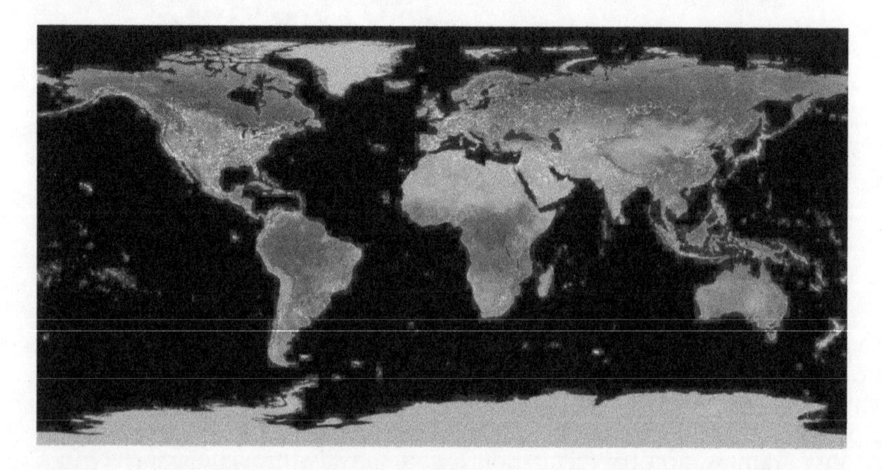

8.11. Map of the earth with a 6-meter (20-ft) sea level rise represented in red. See full-color version in center of book.

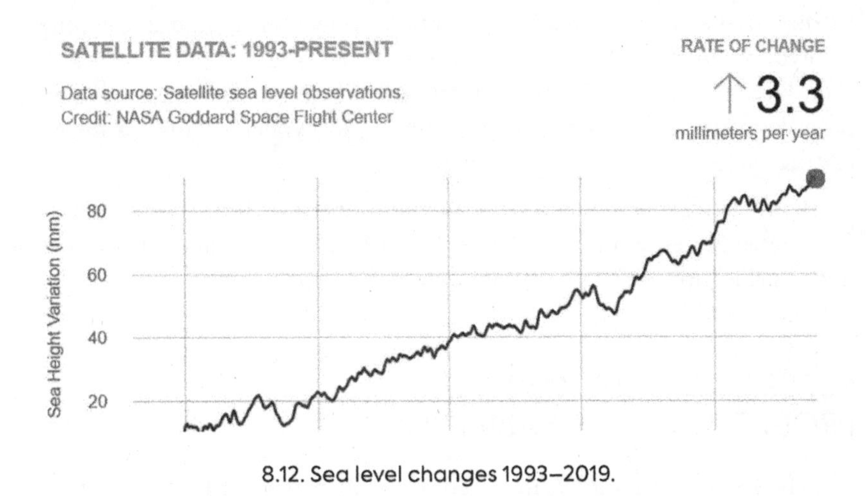

8.12. Sea level changes 1993–2019.

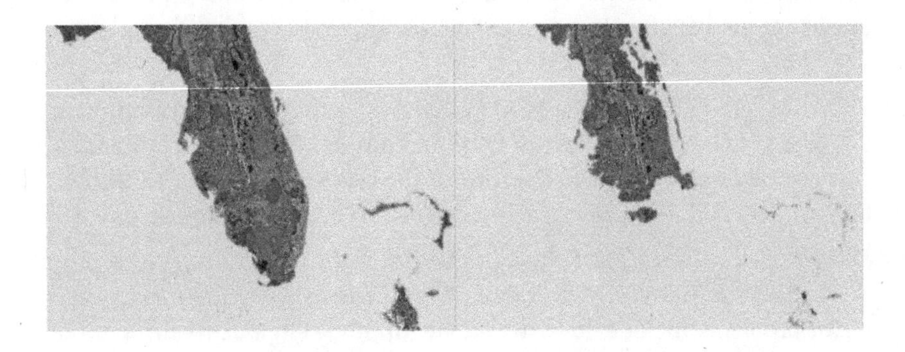

8.13. Map of the US state of Florida showing areas above water after the 3-meter (10-ft) rise predicted 2050–2075. See full-color version in center of book.

WEAPONIZED AI

While shooting wars between countries are rare, code wars rage constantly. Sophisticated countries and organizations use artificial intelligence (AI) to disrupt the electronic systems of rivals. Dams, electrical grids, defense systems, satellites, the Internet, supply chains, and factories are all vulnerable to cyberattacks from smart AI.

Futurist Daniel Schmachtenberger uses the term "weaponized AI" for this capability. He points out that no country can afford to lay down its digital arsenal unilaterally. If Country A decides to stop developing AI cyberweapons, it's quickly overtaken by Country B. Country B then has the ability to bring Country A to its knees. And vice versa. So both countries are forced to keep investing in cyberwarfare in a never-ending cycle.

In an escalating digital arms race, the country with the best AI weapons exercises hegemony over the rest. With the code war invisible and unstoppable, how do you negotiate a global code war armistice?

SUPERSIZING HUMANITY'S PROBLEM-SOLVING ABILITY

The finite earth and weaponized AI are just two of the thorny problems facing humankind. There are many others, natural and human caused. At over $200 trillion, global debt is unpayable; it's three times the size of the entire global economy.

A United Nations analysis found over 1 million species of animals and plants to be at risk of extinction. An endless Middle Eastern war has cost the US $6 trillion since the turn of the century. More than 2 billion human beings don't have access to clean drinking water. In the past 50 years, 29% of the birds in North America have disappeared.

According to a World Economic Forum survey of Millennials, other major problems include government corruption, religious rivalries, food quality, corporate transparency, discrimination, and income inequality. Humanity's list of challenges is long and daunting.

We need the problem-solving genius of Bliss Brain to solve these wicked problems. Caveman Brain produces only rivalry, survivalism, gridlock, inertia, suspicion, and hopelessness.

In Chapter 4 we noted the 490% increase in problem-solving ability that Bliss Brain produces. Our species is much more likely to survive using this quintupling of capacity than more of the Caveman Brain mindset that has taken us to the verge of extinction.

In my book *Mind to Matter* I devote most of a chapter to the branch of science called "emergence." This word refers to the way order arises spontaneously in complex systems.

8.14. In complex systems like flocks of birds, schools of fish, and colonies of ants, order emerges spontaneously.

When you insert Bliss Brain into emergent systems like the earth's industry, or climate, or weaponized AI, or income inequality, you disrupt the system. You bias it toward compassion, and the ability to see the needs of others as equal to your own. To love.

While planetary resources are finite, compassion and creativity are infinite. While we can't stop rivalrous countries from weaponizing AI, spending recklessly, or exterminating species, we can change the consciousness that makes such behavior possible. Bliss Brain may be the only path by which the human race can extricate itself and the planet from disaster.

EMOTIONAL CONTAGION

In Chapter 2 of my book *Mind to Matter,* I review the extraordinary phenomenon called "emotional contagion." Emotions spread through communities just like the flu. A single happy person makes the people around them happier by 34%. Those affected in turn make others happier—by 15%. The ripple keeps spreading, with the next layer happier by 6%.

Do negative people spread contagion too? Yes, but Negative Nancy has less clout than Positive Pete. Having an unhappy connection increases your chances of unhappiness by just 7%.

In-person contact isn't required to induce emotional contagion. We're being affected by our virtual environments too. In a notorious experiment, researchers used an automated system to tweak the emotional content in the timelines of Facebook users. They were able to successfully induce emotional contagion.

When the experimenters manipulated timelines to emphasize positive emotions, users passed them along in their posts and likes. When timelines were skewed toward negative emotions, user posts became gloomier. Within just 7 days, emotional contagion had spread to 689,003 users—no physical or verbal contact required.

The 14% of the US population who are meditating, and the large numbers doing so worldwide, produce positive emotional contagion in many more. Bliss Brain is contagious.

FIELD EFFECTS

Emotional contagion is just one explanation for the growth of meditation. Another is field effects. Everything begins as energy, then works its way into matter. Though energy fields are invisible, they shape matter. Albert Einstein said that, "The field is the sole governing agency of the particle."

Many studies show that human beings are influenced by the energy fields of others. In a series of 148 1-minute trials involving 25 people, trained volunteers going into heart coherence were able to induce coherence in test subjects at a distance. They didn't have to touch their targets to produce the effect. Their energy fields were sufficient.

When you are in a heart coherent state, your heart radiates a coherent electromagnetic signal into the environment around you. This field is detectable by a magnetometer several meters away. When other people enter that coherent energy field, their heart coherence increases too, producing a group field effect.

Not only are we affected by the fields of other people; we're affected by the energies of the planet and solar system. A remarkable series of experiments, conducted by a research team led by Rollin McCraty, director of research at the HeartMath Institute, has linked individual human energy to solar cycles.

McCraty and his colleagues track solar activity using large magnetometers placed at strategic locations on the earth's surface. Solar flares affect the electromagnetic fields of the planet. The researchers compare the ebbs and flows of solar energy with the heart coherence readings of trained volunteers. They have found that when people are in heart coherence, their electromagnetic patterns track those of the solar system.

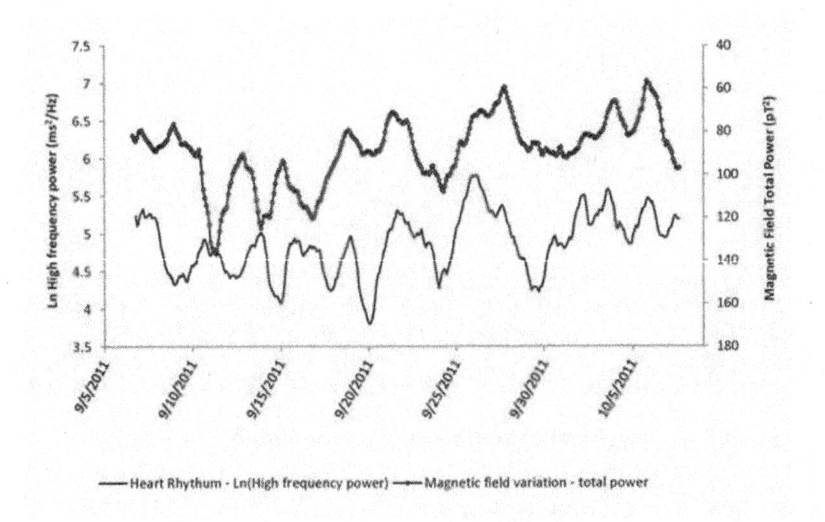

8.15. The heart coherence rhythms of a volunteer compared to solar activity over the course of a month. A later study of 16 participants over 5 months found a similar effect.

McCraty writes: "A growing body of evidence suggests that an energetic field is formed among individuals in groups through which communication among all the group members occurs simultaneously. In other words, there is an actual 'group field' that connects all the members" together.

The results of this research confirm a hypothesis McCraty and I discussed at a conference when I was writing *Mind to Matter*: Not only are these heart-coherent people in sync with large-scale global cycles, they're also *in sync with each other.*

McCraty continues, "We're all like little cells in the bigger Earth brain—sharing information at a subtle, unseen level that exists between all living systems, not just humans, but animals, trees, and so on." When we use selective attention to tune ourselves to positive coherent energy, we participate in the group energy field of other human beings doing the same. We may also resonate in phase with coherent planetary and universal fields.

8.16. The brain functions as receiver of information from the field.

The Brain's Ability to Detect Fields

The idea of invisible energy fields has always been difficult for many scientists to swallow. Around 1900, when Dutch physician Willem Einthoven proposed that the human heart had an energy field, he was ridiculed. He built progressively more sensitive galvanometers to detect it, and he was eventually successful.

Even today, skeptics scoff at field experiments such as those of McCraty. Yet as the evidence continues to mount, it becomes harder to dismiss.

A team of scientists at the California Institute of Technology (Caltech) designed an ingenious experiment to determine if humans could detect energy fields similar to those of the earth.

They hooked participants up to EEGs and confined them in a shielded room, screening out virtually all known sources of energy and radiation. They created a magnetic field generator that precisely mimicked the earth's field.

They then varied the direction of the magnetic field unpredictably, in very short bursts of one-tenth of a second. That's too quick to be consciously detectable. The EEG recorded changes in brain wave amplitudes and frequencies throughout the experiment, which was repeated up to 100 times per subject.

The investigators found drops in alpha waves of up to 60% whenever they changed the direction of the field. They conclude that "the human brain can detect Earth-strength magnetic fields, demonstrating that we have a sensory system that processes the geomagnetic field all around us."

The Caltech authors also noted: "We've known about the five basic senses: vision, hearing, touch, smell, and taste since ancient times, but this is the first discovery of an entirely new human sense in modern times."

JUMP TIME FOR THE PLANET

Through field effects, emotional contagion, social influence, resonance, and the simple human desire to feel good, meditation is spreading. Large and growing numbers of people worldwide are entering Bliss Brain. This accelerates the positive trends toward human thriving we examined in this chapter. It also provides the substrate of compassion that turns consciousness away from Armageddon.

At a Gates Foundation conference, former US president Barack Obama declared, "If you had to choose one moment in history in which to be born, and you didn't know in advance whether you were going to be male or female, which country you were going to be from, what your status was, you'd choose right now." He observes that the world has never been "healthier, or wealthier, or better educated, or in many ways more tolerant, or less violent, than it is today."

As a species, we're moving far beyond the survival mentality of Caveman Brain. We're leaving behind the standards of behavior that defined "normal" in the last century. A critical mass of people is using the human superpower—unique in evolutionary history—to reshape the tissue of their own brains. Bliss Brain is a wonderful-feeling state, but when practiced consistently, it leads to trait change, as neural pathways are repatterned in much healthier ways.

This isn't simply helping us feel better as individuals. It's contributing to Jump Time in collective planetary evolution. Just as the Renaissance of the 1300s changed art, law, education, politics, religion, agriculture, science, and every other facet of human existence, the compassion produced by Bliss Brain transforms the material reality in which we live. This is the most exciting time in all of history to be alive.

As we as a species jump to the next level of flourishing, we are unlocking creative potential the world has never known before. From changing our minds to changing our brains to changing our societies to solving global problems, we're ushering in a completely different future for the planet.

DEEPENING PRACTICES

Here are practices you can do this week to integrate the information in this chapter into your life:

- **Selective Attention Exercise 1:** In what areas of your life do you focus on the negative rather than the positive? Write down three positive affirmations about that area of your life. Make 10 copies. Place one in your wallet. Tape others to your refrigerator, bathroom mirror, computer monitor, video screen, car dashboard, and other places you can't avoid noticing them. Practice repeating the positive affirmations the second you catch yourself focusing on the negative.
- **Journaling Exercise:** Write down a list of personality flaws that you'd like to change. Create a reminder in your online calendar for 1 year from now, reminding you to check today's date in your journal. Next year, you might be surprised to see how much some have shifted after a year of meditation.
- **Emotional Contagion Practice:** Put the power of emotional contagion to work for you. Make a list of the happiest people you know, and make a plan to get together with at least four of them in the coming month.
- **Selective Attention Exercise 2:** Whenever you hear a bad news story that upsets you, do a web search for contradictory evidence (e.g., "Good news about . . ."). This will put the bad news in context.

- **Field Effects Exercise:** Look at the Insight Timer app each day you meditate and notice how many other people are meditating world-wide. It's usually hundreds of thousands. This reminds you that you are not alone.

EXTENDED PLAY RESOURCES

The Extended Play version of this chapter includes:

- Dawson Guided EcoMeditation: Meditating in the Group Energy Field
- The Tide of Grace Chant with Julia and Tyler
- Timeline of the Evolution of Life
- Global Coherence Initiative Video from HeartMath
- Obama's "Greatest Time to Be Alive" Speech
- Countries and Economies in which People Live at Elevations of Less Than 5 Meters
- The Caltech Report on the Field Study

Get the extended play resources at: BlissBrainBook.com/8.

AFTERWORD

The Happiness Habit

I am so grateful to you for joining me on this journey of exploration. I am grateful for the magnificent community of Bliss Brainers of which you and I are a part. As I watch people like us release the suffering that has held us back our whole lives, I feel deeply grateful that we now have the science, the knowledge, the intention, and the practical tools to make this liberation a reality.

My dream is to see these tools become available to everyone. The revolution I describe in Chapter 8 transforms the life of every citizen of the world. It also creates a positive future for other species and the planet itself. My days are spent playing and working toward this goal.

THE PRIORITY THAT TRANSFORMS ALL OTHER PRIORITIES

That revolution begins with consciousness. Transform your consciousness, direct your attention wisely, and the transformation of your material circumstances immediately begins.

According to most traditions, the time when meditation is most potent is in the morning. The delicious state of Bliss Brain is like a broadcast channel. You can decide to tune your awareness to it anytime. When you make this choice first thing in the morning, when your brain is in high alpha and you're moving from sleep to wakefulness, you align your experience with the energy of this channel as the first act of your day.

This is a powerful statement of intention. You are saying to the universe, "My first priority today is to align with you. I choose to live in synchrony with nature and the cosmos the whole of this day, starting now. Nothing is more important to me than this alignment. I surrender my local mind to the greatness of nonlocal mind, and I open the whole of my existence to love, joy, and peace."

Get this priority straight, and all your other priorities line up behind it. By getting in tune with the universe, you step into synchrony with all of nature. You enter a flow state, your life becomes easier, and the challenges you face are placed in the context of the love, joy, and peace found in Bliss Brain.

When I teach live workshops, or give keynote speeches, I often end by asking people to raise their hands if they will commit to meditating for 15 minutes every morning. I've given you links to many free EcoMeditation tracks in this book, and research shows that if you use them regularly, they take you to Bliss Brain.

If you commit to that daily 15 minutes, please raise your hand now.

Thank you! You've just voted yourself into the Bliss Brain community.

A.1. People raising their hands at a live event.

REASONS FOR SUFFERING

I want you to have a happy life. No matter how good your reasons for suffering, most are figments of your imagination. They are invisible chains that can bind you to feeling bad for life. You can say to yourself, "I suffer for this reason _______________ (insert your favorite one here)."

I could still be saying, "I'm suffering because I lost my home and all my possessions in the fire. I'm suffering because I lost my life savings in the financial disaster that followed. I'm suffering because my body broke down when I dragged those heavy boxes around."

But I'm not suffering, despite those extreme events. I made different choices, starting with that moment in the hotel room by the coast when, bewildered and disoriented, I said to Christine, "We need to do something urgently. We need to meditate."

I chose a new story, and turned the tragedy of Chapter 1 into the post-traumatic growth of Chapter 7.

We've all had tragedies in our lives. You've had tragedies in yours. What insults still run riot in your Default Mode Network, transporting the misery of your past into the promise of your future? Cementing the suffering of yesterday into the mystery of tomorrow? Guaranteeing that you suffer subsequently the way you suffered previously?

I invite you to examine every old suffering story of your entire life, and open your mind to the possibility of a new narrative.

We can't change the past, when miserable things happened to us. But we can *change our story about the past.* This exercise aligns us with the power of possibility; we embrace redemption and growth.

Changing our stories doesn't mean that we justify the actions of the people who hurt us. We don't need to forgive till we're 100% ready. And our forgiveness doesn't excuse what they did to us. What it does accomplish is to release our own stress. We're not changing our story to help *them.* We're doing it to help *ourselves,* and liberate our own future from the suffering of the past. While we can't change the past, we can change the story we tell ourselves about the past. That creates a new future.

MAKING BLISS BRAIN A HABIT

I want Bliss Brain to become a habit for you, as it is for the One Percent. Once you experience the neurochemicals of bliss I describe in Chapter 5, and they start to condition your brain, you'll be hooked for life. Within 8 weeks you'll build the neural circuits to regulate your negative emotions and control your attention, as we saw in Chapter 6. You'll turn on the Enlightenment Circuit and downgrade the suffering of Selfing. Within a few months you'll have created the brain hardware of resilience, creativity, and joy.

You'll transform feeling good from a state to a trait. Then, Bliss Brain isn't just *how you feel.* Bliss Brain is *who you are.* Bliss Brain has become your nature, hardwired into the circuits of the four lobes of your brain. It has become your possession, and one so precious that you would never give it up. No one can ever take it away from you.

PERSPECTIVE ON LOCAL LIFE

When you flip the switch into Bliss Brain in meditation each day, you find yourself in a place of infinite peace and joy. You're in a place of pure consciousness. You're not limited by your body or your history.

Experiencing this state feels like the only thing that really matters in life. Local life and local mind have meaning and purpose only when they're lived from this place of nonlocal mind. Daily morning meditation is what anchors you to the experience of infinite awareness.

All the rest of your life is then lived from that place of connection with nonlocal mind. It frames everything, putting local reality into perspective.

All the things that seem so important when you're trapped inside the limits of a local mind seem trivial: money, fame, sex, admiration, opinions, body image, deadlines, goals, achievements, failures, problems, solutions, needs, routines, self-talk, physical ailments, the state of the world, comfort, insults, impulses, discomfort, memories, thoughts, desires, frustrations, plans, timelines, tragedies, events, news, sickness, entertainment, emotions, hurts, games, wounds, compliments, wants, pains, aspirations, past, future, worries, disappointment, urgent items, and demands for your time and attention.

All these things fade into insignificance. All that remains is consciousness. The vast universal now, infused with perfection.

This becomes the perspective from which you view your local life. It's the starting point for each day. It becomes the origination point for everything you think and do that day. Your local reality is shaped by nonlocal mind. You are everything. You have everything. You lack nothing. You proceed into your day, creating from this anchor of perfection. What you create reflects this perfection.

9.2. When in local mind, so many things seem important.

BROADCASTING RESONANCE

Anchored in nonlocal consciousness, your local life begins to change. As you resonate with the cycles of nature, as your heart's coherence conditions the energy space around you, as you vibrate to the signal of love and joy in your consciousness, you attract people and conditions that match your states and traits.

Without effort, as your magnificent new signal broadcasts out around you, resonating with the music of the universe, you'll come into synchrony with people and events that bless and delight you.

You'll discover that you're not alone. As you tune to the great symphony of life each day, you'll find that you're tuned to millions of other people who are likewise attuned. With no effort at all, you'll discover wonderful new friends and companions wherever you travel. As the light shines from your eyes, it meets the light in the eyes of others. When you're awake, you naturally enjoy others who are awake.

9.3. Coming into synchrony.

LOVING THE SLEEPER

Not everyone is awake, and that's fine. Sometimes your friends and family members are tossing in their sleep, suffering unnecessarily. Their plight touches you. You feel their misery. You would love to see them wake up, and shed those beliefs, thoughts, and habits that drag them down.

You can't force them to do so, no matter how much you love them. Everyone makes their own choice.

What you *can* do for people who are suffering is shine brightly yourself. If they're ready, they'll wake up. If they don't, trust the universe. We each wake up when the time is right. Their time might come later; it's not up to you. You can share this book and other resources with them. You can share your story as I have shared mine, and perhaps these examples will inspire them. If and when each of us wakes up is our choice.

UNLOCKING YOUR POTENTIAL

As you live in synchrony with the universe, enjoying the community of other Bliss Brainers, you find new possibilities opening up. You start to unlock potential that's been trapped inside the suffering, selfing self.

Increasingly, you're not just in Bliss Brain during meditation. You're in the Awakened Mind state with your eyes open, going about your day. All kinds of possibilities that were previously unavailable to you now become available.

For 25 years, I worked in book publishing. I wasn't an author; my job was to promote the work of *important* people—people with something to say—by managing the publication of their books. I didn't think I had anything valuable to say to the world, or any great contribution to make to human awakening. I was content to be an administrator, working behind the scenes.

When I began to meditate daily, that started to change. I discovered that I had a voice, and that when I gave a presentation, people were inspired. I realized the importance of science, and even though I was nearly 50 years old, I retrained myself as a researcher. I co-authored the book *Soul Medicine,* about the power of energy therapies to heal.

I became aware that when science and mysticism were brought together, they formed a potent transformational combination. I realized that meditation and energy work could produce huge shifts in gene expression, and that science could measure these changes. Science could also show us which spiritual practices were effective, and which were window dressing.

In my next book, *The Genie in Your Genes,* I wrote about these discoveries, and people responded. This launched a completely new career path. It embodied potential I never before realized I had.

Imagine similar shifts happening for you.

Expect your life to become different. Now that you've raised your hand, expect Bliss Brain to take you far beyond where you begin today. Expect your love life, your career, your family relationships, your physical body, your money, your friendships, your spiritual path, and your sense of well-being to be utterly transformed.

Expect your potential to be unleashed. Expect yourself to taste ecstasy every day. Expect flow states to become your new normal. Expect elevated emotions to course through your heart while inspired thoughts flood your mind. Expect adversity to strengthen rather than crush you. Expect your days to begin and end with bliss. Expect to be a happy person. Expect to do things you never believed possible.

My eyes are full of tears as I write this, my love letter to you. I have poured my heart, mind, and soul into writing this book, with the goal of inspiring you to claim your full potential. Now it's your turn. It's time to put this all into practice as you create an extraordinary life for yourself. I'll see you on the journey.

REFERENCES

Chapter 2

A 14th-century Tibetan mystic: Rinpoche, T. D. (2008). *Great perfection, volume II: Separation and breakthrough.* Ithaca, NY: Snow Lion.

People often become much more spiritual after mystical breakthroughs: Newberg, A., & Waldman, M. R. (2017). *How enlightenment changes your brain: The new science of transformation.* New York, NY: Penguin.

Greeley studied nonordinary states: Greeley, A. (1974). *Ecstasy: A way of knowing.* New Jersey: Prentice Hall.

Elements common to transcendent experience: Greeley, A. (1975). *The sociology of the paranormal: A reconnaissance.* Thousand Oaks, CA: Sage.

Later researchers built on Greeley's initial findings: Spilka, B., Brown, G. A., & Cassidy, S. A. (1992). The structure of religious mystical experience in relation to pre- and postexperience lifestyles. *International Journal for the Psychology of Religion, 2*(4), 241–257.

Huxley's sacramental vision of reality and other quotes: Huxley, A. (1954). *The doors of perception.* London: Chatto & Windus.

William James's statements after his own transcendent experiences: James, W. (1902). *The varieties of religious experience.* NY: Longman.

The brains of the mystics share a common profile: Newberg, A., & Waldman, M. R. (2017). *How enlightenment changes your brain: The new science of transformation.* New York, NY: Penguin.

An important study performed at Emory University: Hasenkamp, W., & Barsalou, L. W. (2012). Effects of meditation experience on functional connectivity of distributed brain networks. *Frontiers in Human Neuroscience, 6,* 38.

Francis de Sales quotes: Huxley, A. (1947). *The perennial philosophy* (p. 285). New York, NY: Harper & Row.

Over 90% report that enlightenment experiences are more real than everyday reality: Newberg, A., & Waldman, M. R. (2017). *How enlightenment changes your brain: The new science of transformation.* New York, NY: Penguin.

Byron Katie: Massad, S. (2001). An interview with Byron Katie. *Realization.* Retrieved December 1, 2018, from https://realization.org/p/byron-katie/massad.byron-katie-interview/massad.byron-katie-interview.1.html

The brain's energy usage rarely varies more than 5%: Newberg, A., & Waldman, M. R. (2017). *How enlightenment changes your brain: The new science of transformation.* New York, NY: Penguin.

The Task-Positive Network or TPN: Mulders, P. C., van Eijndhoven, P. F., & Beckmann,

C. F. (2016). Identifying large-scale neural networks using fMRI. In *Systems neuroscience in depression* (pp. 209–237). Academic Press.

In describing this discovery, Raichle notes: Raichle, M. E. (2014). The restless brain: how intrinsic activity organizes brain function. *Philosophical Transactions of the Royal Society of London. Series B, Biological Sciences, 370*(1668), 20140172. doi:10.1098/rstb.2014.0172

When we relax, the Default Mode Network is the most active area of the brain: Davanger, S. (2015, March 9). *The brain's default mode network—what does it mean to us? The Meditation Blog.* Retrieved November 13, 2018, from https://www.themeditationblog.com/the-brains-default-mode-network-what-does-it-mean-to-us

Raichle theorizes that such activity helps the brain stay organized: Hughes, V. (2010, October 6). *The brain's dark energy. National Geographic.* Retrieved November 13, 2018, from https://www.nationalgeographic.com/science/phenomena/2010/10/06/brain-default-mode

The DMN recruits other brain areas: Vatansever, D., Manktelow, A. E., Sahakian, B. J., Menon, D. K., & Stamatakis, E. A. (2017). Angular default mode network connectivity across working memory load. *Human Brain Mapping, 38*(1), 41–52.

The wandering mind of the DMN has a "me" orientation: Mason, M. F., Norton, M. I., Van Horn, J. D., Wegner, D. M., Grafton, S. T., & Macrae, C. N. (2007). Wandering minds: The default network and stimulus-independent thought. *Science, 315*(5810), 393–395.

The demon when Buddha was on the verge of enlightenment: Gyatso, K. (2008). *Introduction to Buddhism: An explanation of the Buddhist way of life.* Glen Spey, NY: Tharpa.

This regulatory neural connection is stronger in adept meditators: Froeliger, B., Garland, E. L., Kozink, R. V., Modlin, L. A., Chen, N. K., McClernon, F. J., . . . Sobin, P. (2012). Meditation-state functional connectivity (msFC): strengthening of the dorsal attention network and beyond. *Evidence-Based Complementary and Alternative Medicine, 2012,* 1–10. doi:10.1155/2012/680407

The nucleus accumbens shrinks in longtime meditators: Marchand, W. R. (2014). Neural mechanisms of mindfulness and meditation: evidence from neuroimaging studies. *World Journal of Radiology, 6*(7), 471.

People spend about 47% of their time in negativity: Bradt, S. (2010). Wandering mind not a happy mind. *Harvard Gazette.* Retrieved November 15, 2018, from https://news.harvard.edu/gazette/story/2010/11/wandering-mind-not-a-happy-mind

People were at their happiest when their attention was focused in the present moment: Killingsworth, M. A., & Gilbert, D. T. (2010). A wandering mind is an unhappy mind. *Science, 330*(6006), 932. doi:10.1126/science.1192439

Meditators can shut off the DMN: Brewer, J. A., Worhunsky, P. D., Gray, J. R., Tang, Y. Y., Weber, J., & Kober, H. (2011). Meditation experience is associated with differences in default mode network activity and connectivity. *Proceedings of the National Academy of Sciences, 108*(50), 20254-20259.

The DMN recruits other brain regions: Fox, K. C., Spreng, R. N., Ellamil, M., Andrews-Hanna, J. R., & Christoff, K. (2015). The wandering brain: Meta-analysis of functional neuroimaging studies of mind-wandering and related spontaneous thought processes. *Neuroimage, 111,* 611–621.

Mark Leary and the curse of self: Leary, M. R. (2007). *The curse of the self: Self-awareness, egotism, and the quality of human life.* Oxford, UK: Oxford University Press.

The DMN is active in self-oriented and social tasks: Sormaz, M., Murphy, C., Wang, H. T., Hymers, M., Karapanagiotidis, T., Poerio, G., . . . Smallwood, J. (2018). Default mode network can support the level of detail in experience during active task states. *Proceedings of the National Academy of Sciences, 115*(37), 9318–9323.

Kegan and the subject-object shift: Kegan, R. (1994). *In over our heads: The mental demands of modern life.* Cambridge, MA: Harvard University Press.

Our emotions and thoughts become less "sticky" and lose their self-hypnotic power: Goleman, D., & Davidson, R. J. (2018). *Altered traits: Science reveals how meditation changes your mind, brain, and body.* New York: Avery.

American spending on illicit drugs and alcohol: Midgette, G., Davenport, S., Caulkins, J. P., & Kilmer, B. (2019). *What America's Users Spend on Illegal Drugs, 2006–2016.* Santa Monica, CA: RAND.

Four-trillion-dollar Altered States Economy: Kotler, S., & Wheal, J. (2017). *Stealing fire: How Silicon Valley, the Navy SEALs, and maverick scientists are revolutionizing the way we live and work* (p. 28). New York, NY: HarperCollins.

The human drive for ecstasy as far back in time as the ancient Greeks: Kotler, S., & Wheal, J. (2017). *Stealing fire: How Silicon Valley, the Navy SEALs, and maverick scientists are revolutionizing the way we live and work.* New York, NY: HarperCollins.

Ronald Siegel "fourth drive": Siegel, R. K. (1998). *Intoxication: The universal drive for mind-altering substances* (p. 11). Rochester, VT: Park Street.

Meditators are able to deactivate the brain's posterior cingulate cortex: Van Lutterveld, R., Houlihan, S. D., Pal, P., Sacchet, M. D., McFarlane-Blake, C., Patel, P. R., . . . Brewer, J. A. (2017). Source-space EEG neurofeedback links subjective experience with brain activity during effortless awareness meditation. *NeuroImage, 151,* 117–127.

Studies of Tibetan monks with tens of thousands of hours of practice: Brewer, J. A., Worhunsky, P. D., Gray, J. R., Tang, Y. Y., Weber, J., & Kober, H. (2011). Meditation experience is associated with differences in default mode network activity and connectivity. *Proceedings of the National Academy of Sciences, 108*(50), 20254–20259.

The brains of the mystics making the subject-object shift share a common profile: Newberg, A., & Waldman, M. R. (2017). *How enlightenment changes your brain: The new science of transformation.* New York, NY: Penguin.

Regions associated with Bliss Brain: Tang, Y. Y., Hölzel, B. K., & Posner, M. I. (2015). The neuroscience of mindfulness meditation. *Nature Reviews Neuroscience, 16*(4), 213.

Chapter 3

Function of brain regions: Corballis, M. C. (2014). Left brain, right brain: Facts and fantasies. *PLoS Biology, 12*(1). doi:10.1371/journal.pbio.1001767

The three part brain: Audesirk, T., Audesirk, G., & Byers, B. E. (2008). *Biology: Life on earth with physiology.* Upper Saddle River, NJ: Pearson Prentice Hall.

In master meditators, brain activity drops up to 40%: Newberg, A., & Waldman, M. R. (2017). *How enlightenment changes your brain: The new science of transformation.* New York, NY: Penguin.

Loneliness and isolation have detrimental effects on our health: Holt-Lunstad, J., Smith, T. B., & Layton, J. B. (2010). Social relationships and mortality risk: A meta-analytic review. *PLoS Medicine, 7*(7), e1000316.

A Stanford study of people who experienced the deep now: Rudd, M., Vohs, K., & Aakers, J. (2011). Awe expands people's perception of time, alters decision-making, and enhances well-being. *Psychological Science, 23*(10), 1130-1136.

Jill Bolte Taylor's stroke: Koontz, K. (2018). *Listening in with Jill Bolte Taylor. Unity,* January/February. Retrieved December 9, 2019, from http://www.unity.org/publications/unity-magazine/articles/listening-jill-bolte-taylor

Balance between the two hemispheres: Corballis, M. C. (2014). Left brain, right brain: Facts and fantasies. *PLoS Biology, 12*(1). doi:10.1371/journal.pbio.1001767

The corpus callosum in meditators: Luders, E., Kurth, F., Mayer, E. A., Toga, A. W., Narr, K. L., & Gaser, C. (2012). The unique brain anatomy of meditation practitioners: Alterations in cortical gyrification. *Frontiers in Human Neuroscience, 6,* 34.

Babe Ruth's senses were more acute than average: Fullerton, H. S. (1921). Why Babe Ruth is the greatest home-run hitter. *Popular Science, 99*(4), 19–21.

Einstein's brain had greater volume in the corpus callosum: Men, W., Falk, D., Sun, T., Chen, W., Li, J., Yin, D., . . . Fan, M. (2014). The corpus callosum of Albert Einstein's brain: Another clue to his high intelligence? *Brain, 137*(4), e268-e278.

Hippocampus functions: Chan, R. W., Leong, A. T., Ho, L. C., Gao, P. P., Wong, E. C., Dong, C. M., . . . Wu, E. X. (2017). Low-frequency hippocampal–cortical activity drives brain-wide resting-state functional MRI connectivity. *Proceedings of the National Academy of Sciences, 114*(33), E6972–E6981.

The short circuit improves reaction speed, but at the expense of accuracy: Sapolsky, R. M. (2017). *Behave: The biology of humans at our best and worst.* New York, NY: Penguin.

The short and the long path: Church, D. (2017). *Psychological trauma: Healing its roots in brain, body, and memory.* Santa Rosa, CA: Energy Psychology Press.

Amygdala functions: Church, D. (2017). *Psychological trauma: Healing its roots in brain, body, and memory.* Santa Rosa, CA: Energy Psychology Press.

The difference in amygdala activation is 400%: Goleman, D., & Davidson, R. J. (2018). *Altered traits: Science reveals how meditation changes your mind, brain, and body.* New York, NY: Avery.

Von Economo neurons facilitate the integration of information: Namkung, H., Kim, S. H., & Sawa, A. (2017). The insula: An underestimated brain area in clinical neuroscience, psychiatry, and neurology. *Trends in Neurosciences, 40*(4), 200–207.

Vietnam veterans and social engagement: Lutz, A., Brefczynski-Lewis, J., Johnstone, T., & Davidson, R. J. (2008). Regulation of the neural circuitry of emotion by compassion meditation: Effects of meditative expertise. *PloS One, 3*(3), e1897.

The TPJs of meditators process efficiently after 40 days: Hölzel, B. K., Carmody, J., Vangel, M., Congleton, C., Yerramsetti, S. M., Gard, T., & Lazar, S. W. (2011). Mindfulness practice leads to increases in regional brain gray matter density. *Psychiatry Research: Neuroimaging, 191*(1), 36–43.

Olaf Blanke's out-of-body patient: Blakeslee, S. (2006, October 3). Out of body experience? Your brain is to blame. *The New York Times.* http://www.nytimes.com/2006/10/03/health/psychology/03shad.html

Insula functions: Blakeslee, S. (2007, February 6). A small part of the brain, and its profound effects. *The New York Times.* https://www.nytimes.com/2007/02/06/health/psychology/06brain.html

The level of activity in Mingyur's empathy circuits rose by 700%: Goleman, D., & Davidson, R. J. (2018). *Altered traits: Science reveals how meditation changes your mind, brain, and body.* New York, NY: Avery.

The habitual reified dualities between subject and object, self and other, in-group and out-group dissipate: Vieten, C., Wahbeh, H., Cahn, B. R., MacLean, K., Estrada, M., Mills, P., . . . Presti, D. E. (2018). Future directions in meditation research: Recommendations for expanding the field of contemplative science. *PloS One, 13*(11), e0205740.

Prefrontal cortex changes with age: U.S. Department of Health and Human Services. (n.d.). Maturation of the prefrontal cortex. Retrieved December 9, 2019, from http://www.hhs.gov/opa/familylife/tech_assistance/etraining/adolescent_brain/Development/prefrontal_cortex

Ventromedial prefrontal cortex functions: Viviani, R. (2014). Neural correlates of emotion regulation in the ventral prefrontal cortex and the encoding of subjective value and economic utility. *Frontiers in Psychiatry, 5,* 123.

Dorsolateral prefrontal cortex functions: Miller, E. K., & Buschman, T. J. (2013). Cortical circuits for the control of attention. *Current Opinion in Neurobiology, 23*(2), 216–222.

Once the dorsolateral PFC has made a decision, it suppresses incoming information that is now irrelevant: Sapolsky, R. M. (2017). *Behave: The biology of humans at our best and worst.* New York, NY: Penguin.

Students trained in mindfulness improved their scores on the GRE: Mrazek, M. D., Franklin, M. S., Phillips, D. T., Baird, B., & Schooler, J. W. (2013). Mindfulness training improves working memory capacity and GRE performance while reducing mind wandering. *Psychological Science, 24*(5), 776–781.

Caudate nucleus helps us remember what we've learned: Fox, K. C., Dixon, M. L., Nijeboer, S., Girn, M., Floman, J. L., Lifshitz, M., . . . Christoff, K. (2016). Functional neuroanatomy of meditation: A review and meta-analysis of 78 functional neuroimaging investigations. *Neuroscience and Biobehavioral Reviews, 65,* 208–228.

High performers engage the caudate nucleus: Vago, D. R. (2015, July 31). Brain's response to meditation: How much meditation does it take to change your brain and relieve stress? *Psychology Today.* https://www.psychologytoday.com/us/blog/the-science-behind-meditation/201507/brains-response-meditation

To the brain, imagination is reality: Reddan, M. C., Wager, T. D., & Schiller, D. (2018). Attenuating neural threat expression with imagination. *Neuron, 100*(4), 994–1005.

Imagination is a neurological reality investigator quotes: University of Colorado at Boulder. (2018, December 10). Your brain on imagination: It's a lot like reality, study shows. *Science Daily.* https://www.sciencedaily.com/releases /2018/12/181210144943.htm

Negative thinking will interrupt the brain's ability to perform well on every level: Newberg, A., & Waldman, M. R. (2017). *How enlightenment changes your brain: The new science of transformation.* New York, NY: Penguin.

This turns on hundreds of genes that regulate our immunity: Bhasin, M. K., Dusek, J. A., Chang, B. H., Joseph, M. G., Denninger, J. W., Fricchione, G. L., . . . Libermann, T. A. (2013). Relaxation response induces temporal transcriptome changes in energy metabolism, insulin secretion and inflammatory pathways. *PLoS One, 8*(5), e62817.

A transformation that dramatically ups the limits: Goleman, D., & Davidson, R. J. (2018). *Altered traits: Science reveals how meditation changes your mind, brain, and body.* New York, NY: Avery.

An overview of meditation studies shows that this opens up nonlocal consciousness: Vieten, C., Wahbeh, H., Cahn, B. R., MacLean, K., Estrada, M., Mills, P., . . . Presti, D. E. (2018). Future directions in meditation research: Recommendations for expanding the field of contemplative science. *PloS one, 13*(11), e0205740.

Chapter 4

The top 1% of the world's wealthy: Credit Suisse Research Institute. (2018). *Global wealth report.* Credit Suisse. Retrieved December 9, 2019, from https://www.credit-suisse.com/corporate/en/research/research-institute/global-wealth-report.html

In the early 1300s, England had a monastic population of about 22,000: Mortimer, I. (2009). *The time traveler's guide to medieval England: A handbook for visitors to the fourteenth century.* New York, NY: Simon and Schuster.

Estimates are similar for other medieval European countries: Brucker, G. A. (1998). *Florence: The golden age, 1138–1737.* Berkeley: University of California Press.

Percentage of monastics in Tibet and Thailand: Goldstein, M. C. (2010). Tibetan Buddhism and mass monasticism. In A. Herrou and G. Krauskopff (Eds.), *Des moines et des moniales de part le monde: La vie monastique dans le miroir de la parenté.* France: Presses Universitaires de Toulouse le Mirail.

Roman historian Cicero wrote: Cicero, M. T. (2017). *On the commonwealth and on the laws.* New York, NY: Cambridge University Press.

The Eleusinian mysteries: Tripolitis, A. (2002). *Religions of the Hellenistic-Roman age.* Grand Rapids, MI: W. B. Eerdmans.

Steps of initiation included: Hoffman, A., Wasson, R. G., & Ruck, C. A. P. (1978). *The Road to Eleusis: Unveiling the Secret of the Mysteries.* New York, NY: Harcourt, Brace, Jovanovich.

Max Cade and the Mind Mirror: Cade, M., & Coxhead, N. (1979). *The awakened mind: Biofeedback and the development of higher states of awareness.* New York, NY: Dell.

Robert Becker and early EEG work: Becker, R. O. (1990). The machine brain and properties of the mind. *Subtle Energies and Energy Medicine Journal Archives, 1*(2).

The Awakened Mind pattern and Anna Wise: Wise, A. (1995). *The high performance mind: Mastering brainwaves for insight, healing, and creativity.* New York, NY: Putnam. Wise, A. (2002). *Awakening the mind: A guide to harnessing the power of your brainwaves.* New York, NY: Tarcher.

People in flow states and big alpha: MindLabPro. (2018, May 12). Nootriopics for flow state—brain-boosters to initiate supreme task performance. Retrieved from https://www.mindlabpro.com/blog/nootropics/nootropics-flow-state

The age of the neocortex and brain stem: Rakic, P. (2009). Evolution of the neocortex: A perspective from developmental biology. *Nature Reviews Neuroscience, 10*(10), 724.

Less than 10% of the US population meditates: Clarke, T. C., Black, L. I., Stussman, B. J., Barnes, P. M., & Nahin, R. L. (2015). Trends in the use of complementary health approaches among adults: United States, 2002–2012. *National Health Statistics Reports, 79,* 1–16.

This is because meditation is hard: Rutrowski, R. (2018). Only 12% of U.S. adults use meditation. Here's why . . . *HerbalMind.* Retrieved December 9, 2019, from https://www.herbalmindlife.com/only-12-of-us-adults-use-meditation-heres-why

Relaxing your tongue on the floor of your mouth: Schmidt, J. E., Carlson, C. R., Usery, A. R., & Quevedo, A. S. (2009). Effects of tongue position on mandibular muscle activity and heart rate function. *Oral Surgery, Oral Medicine, Oral Pathology, Oral Radiology, and Endodontology, 108*(6), 881–888.

Imagining the volume of space inside your body puts you into alpha: Fehmi, L., & Robbins, J. (2008). *The open-focus brain: Harnessing the power of attention to heal mind and body.* Boston, MA: Shambhala/Trumpeter Books.

Sending a beam of energy through your heart: McCraty, R., Atkinson, M., Tomasino, D., & Bradley, R. T. (2009). The coherent heart: Heart-brain interactions, psychophysiological coherence, and the emergence of system-wide order. *Integral Review: A Transdisciplinary and Transcultural Journal for New Thought, Research, and Praxis, 5*(2), 7–14.

Compassion produces brain changes in meditators: Goleman, D., & Davidson, R. J. (2018). *Altered traits: Science reveals how meditation changes your mind, brain, and body.* New York, NY: Avery.

In two days, many participants acquired elevated brain states: Pennington, J., Sabot, D., & Church, D. (2019). EcoMeditation and EFT (Emotional Freedom Techniques) produce elevated brainwave patterns and states of consciousness. *Energy Psychology: Theory, Research, and Treatment, 11*(1), 13-28. *The Bond University fMRI study:* Stapleton, P., Baumann, O., Church, D., & Sabot, D. (2020) Functional brain changes associated with EcoMeditation. Reported at Transformational Leadership Conference, Panama City, Panama, Jan 26. Submitted for publication.

Esalen Institute EcoMeditation study: Groesbeck, G., Bach, D., Stapleton, P., Blickheuser, K., Church, D., & Sims, R. (2017). The interrelated physiological and psychological effects of EcoMeditation. *Journal of Evidence-Based Integrative Medicine, 23,* 1–6. doi:10.1177/2515690X18759626.

The 40% drop in PFC function measured in meditators: Newberg, A., & Waldman, M. R. (2017). *How enlightenment changes your brain: The new science of transformation.* New York, NY: Penguin.

Paul Brunton's experiences: Brunton, P. (2014). *The short path to enlightenment.* Burdett, NY: Larson.

You build muscle much more quickly when you lift weights very slowly: Ferriss, T. (2010). *The four-hour body.* New York, NY: Crown Archetype.

Weight lifting doesn't require more than 30 minutes a session: Ferriss, T. (2010). *The four-hour body.* New York, NY: Crown Archetype.

Different meditation methods produce different effects in the brain: Lumma, A. L., Kok, B. E., & Singer, T. (2015). Is meditation always relaxing? Investigating heart rate, heart rate variability, experienced effort and likeability during training of three types of meditation. *International Journal of Psychophysiology, 97*(1), 38–45.

The 10,000-hour rule: Goleman, D., & Davidson, R. J. (2018). *Altered traits: Science reveals how meditation changes your mind, brain, and body* (p. 258). New York, NY: Avery.

Large-scale study at a workshop taught by Dr. Joe Dispenza: Church, D., Yang, A., Fannin, J., & Blickheuser, K. (2018). *The biological dimensions of transcendent states: A randomized controlled trial.* Presented at the research meeting of the conference of the Association for Comprehensive Energy Psychology (ACEP), May 31, 2018.

When you sustain alpha brain waves, several other waves change: Pennington, J. (2012). *Your psychic soul: Embracing your sixth sense.* Virginia Beach, VA: 4th Dimension Press.

Advanced yogis have 25 times the gamma activity of ordinary people: Goleman, D., & Davidson, R. J. (2018). *Altered traits: Science reveals how meditation changes your mind, brain, and body.* New York, NY: Avery.

Gamma is observed in states of mystical union: Beauregard, M., & Paquette, V. (2008). EEG activity in Carmelite nuns during a mystical experience. *Neuroscience Letters, 444,* 1–4. doi:10.1016/j.neulet.2008.08.028

Gamma synchrony across brain regions: Goleman, D., & Davidson, R. J. (2018). *Altered traits: Science reveals how meditation changes your mind, brain, and body.* New York, NY: Avery.

Gamma is observed in ordinary people having a moment of creative insight: Jung-Beeman, M., Bowden, E. M., Haberman, J., Frymiare, J. L., Arambel-Liu, S., Greenblatt, R., . . . Kounious, J. (2004). Neural activity when people solve problems with insight. *PLoS Biology, 2,* 0500–0510. doi:10.1371/journal.pbio.0020097

Gamma is also associated with:

Feelings of love and compassion: Pennington, J. (2012). *Your psychic soul: Embracing your sixth sense.* Virginia Beach, VA: 4th Dimension Press.

Increased perceptual organization: Elliott, M. A., & Muller, H. J. (1998). Synchronous information presented in 40-Hz flicker enhance visual feature binding. *Psychological Science, 9,* 277–283. doi:10.1111/1467-9280.00055

Associative learning: Miltner, W. H., Braun, C., Arnold, M., Witte, H., & Taub, E. (1999). Coherence of gamma-band EEG activity as a basis for associative learning. *Nature, 397*(6718), 434–436. doi:10.1038/17126

Efficiency of information flowing across synapses: Sheer, D. E. (1975). Biofeedback training of 40-Hz EEG and behavior. In N. Burch & H. I. Altshuler (Eds.), *Behavior and brain electrical activity.* New York, NY: Plenum.

Healing: Hendricks, L., Bengston, W. F., & Gunkelman, J. (2010). The healing connection: EEG harmonics, entrainment, and Schumann's resonances. *Journal of Scientific Exploration, 24,* 655–666.

Attention and states of transcendent bliss: Das, N., & Gastaut, H. (1955). Variations in the electrical activity of the brain, heart, and skeletal muscles during yogic meditation and trance. *Electroencephalography and Clinical Neurophysiology, 6,* 211–219.

Gamma waves synchronize the four lobes of the brain: Engel, A., Konig, P., Kreiter, A., & Singer, W. (1991). Direct physiological evidence for scene segmentation by temporal coding. *Proceedings of the National Academy of Science, 88*(20), 9136–9140. Pennington, J. (2012). *Your psychic soul: Embracing your sixth sense.* Virginia Beach, VA: 4th Dimension Press.

Neurofeedback expert Judith Pennington, studying eight subjects: Pennington, J., Sabot, D., & Church, D. (2019). EcoMeditation and EFT (Emotional Freedom Techniques) produce elevated brainwave patterns and states of consciousness. *Energy Psychology: Theory, Research, and Treatment, 11*(1), 13-28.

Study of 208 participants at a meditation retreat: Church, D., & Sabot, D. (2019). Group-based EcoMeditation is associated with improved anxiety, depression, PTSD, pain and happiness. *Global Advances in Health and Medicine* (in press).

Types of meditation: Bornemann, B., Herbert, B. M., Mehling, W. E., & Singer, T. (2015). Differential changes in self-reported aspects of interoceptive awareness through 3 months of contemplative training. *Frontiers in Psychology, 5,* 1504.

Planck study compared people experiencing empathy with those actively engaging loving-kindness: Klimecki, O. M., Leiberg, S., Lamm, C., & Singer, T. (2012). Functional neural plasticity and associated changes in positive affect after compassion training. *Cerebral Cortex, 23*(7), 1552–1561.

Strong echoes of the brain patterns of experienced meditators were found in novices: Weng, H. Y., Fox, A. S., Shackman, A. J., Stodola, D. E., Caldwell, J. Z., Olson, M. C., . . . Davidson, R. J. (2013). Compassion training alters altruism and neural responses to suffering. *Psychological Science, 24(7), 1171–1180.*

Telomere study quotes comparing a non-meditating group with groups doing mindfulness and loving-kindness meditation: Dolan, E. W. (2019, August 24). Study provides evidence that loving-kindness meditation slows cellular aging. *Psypost.* Retrieved September 10, 2019, from https://www.psypost.org/2019/08/study-provides-evidence-that-loving-kindness-meditation-slows-cellular-aging-54316

Adrenaline and drastically reduced life span: De Rosa, M. J., Veuthey, T., Florman, J., Grant, J., Blanco, M. G., Andersen, N., . . . Alkema, M. J. (2019). The flight response impairs cytoprotective mechanisms by activating the insulin pathway. *Nature, 573,* 135–138.

Initial studies showed that 8 weeks of mindfulness meditation had an effect: Goldin, P. R., & Gross, J. J. (2010). Effects of mindfulness-based stress reduction (MBSR) on emotion regulation in social anxiety disorder. *Emotion, 10*(1), 83–91.

Desbordes, G., Negi, L. T., Pace, T. W., Wallace, B. A., Raison, C. L., & Schwartz, E. L. (2012). Effects of mindful-attention and compassion meditation training on amygdala response to emotional stimuli in an ordinary, non-meditative state. *Frontiers in Human Neuroscience, 6*(292), 1–15.

Even 4 weeks of meditation made a difference: Walsh, E., Eisenlohr-Moul, T., & Baer, R. (2016). Brief mindfulness training reduces salivary IL-6 and TNF-α in young women with depressive symptomatology. *Journal of Consulting and Clinical Psychology, 84*(10), 887–897.

Pace, T. W., Negi, L. T., Adame, D. D., Cole, S. P., Sivilli, T. I., Brown, T. D., . . . Raison, C. L. (2009). Effect of compassion meditation on neuroendocrine, innate immune and behavioral responses to psychosocial stress. *Psychoneuroendocrinology, 34*(1), 87–98.

Even 4 days of meditation were associated with increased cognitive flexibility, creativity, memory, and attention: Zeidan, F., Johnson, S. K., Diamond, B. J., David, Z., & Goolkasian, P. (2010). Mindfulness meditation improves cognition: Evidence of brief mental training. *Consciousness and Cognition, 19*(2), 597–605.

More pervasive brain changes in those that went on long intensive retreats: Sahdra, B. K., MacLean, K. A., Ferrer, E., Shaver, P. R., Rosenberg, E. L., Jacobs, T. L., . . . Mangun, G. R. (2011). Enhanced response inhibition during intensive meditation training predicts improvements in self-reported adaptive socioemotional functioning. *Emotion, 11*(2), 299–312.

Goleman, D., & Davidson, R. J. (2018). *Altered traits: Science reveals how meditation changes your mind, brain, and body.* New York, NY: Avery.

Ten weeks of meditation is enough to move the needle: Lumma, A. L., Kok, B. E., & Singer, T. (2015). Is meditation always relaxing? Investigating heart rate, heart rate variability, experienced effort and likeability during training of three types of meditation. *International Journal of Psychophysiology, 97*(1), 38–45.

400% decrease in amygdala activation: Goleman, D., & Davidson, R. J. (2018). *Altered traits: Science reveals how meditation changes your mind, brain, and body.* New York, NY: Avery.

The three levels of meditation: Goleman, D., & Davidson, R. J. (2018). *Altered traits: Science reveals how meditation changes your mind, brain, and body.* New York, NY: Avery.

Seven minutes of loving-kindness can increase positive mood and social connection: Hutcherson, C. A., Seppala, E. M., & Gross, J. J. (2008). Loving-kindness meditation increases social connectedness. *Emotion, 8*(5), 720.

Survey of 1,120 meditators: Vieten, C., Wahbeh, H., Cahn, B. R., MacLean, K., Estrada, M., Mills, P., . . . Presti, D. E. (2018). Future directions in meditation research: Recommendations for expanding the field of contemplative science. *PloS One, 13*(11), e0205740.

It takes 40 to 60 minutes to achieve a deep contemplative state: Newberg, A., & Waldman, M. R. (2017). *How enlightenment changes your brain: The new science of transformation.* New York, NY: Penguin.

After just the first 8 minutes of mindfulness, attentional focus improves: Goleman, D., & Davidson, R. J. (2018). *Altered traits: Science reveals how meditation changes your mind, brain, and body.* New York, NY: Avery.

Increases in gray matter of brain areas after 27-minute meditation sessions: Lazar, S. (2014). Change in brainstem gray matter concentration following a mindfulness-based intervention is correlated with improvement in psychological well-being. *Frontiers in Human Neuroscience, 8,* 33.

Among Tibetan monks, there were more pervasive brain changes in those that went on long intensive retreats: Goleman, D., & Davidson, R. J. (2018). *Altered traits: Science reveals how meditation changes your mind, brain, and body.* New York, NY: Avery.

When you intensify the experience, the parietal lobe shuts down and selfing ceases: Newberg, A., & Waldman, M. R. (2017). *How enlightenment changes your brain: The new science of transformation.* New York, NY: Penguin.

Lunch with Warren Buffett:

Umoh, R. (2018, May 28). *These two men paid over $650,000 for lunch with Warren Buffett—here are 3 things they learned.* Make It. CNBC.com. https://www.cnbc.com/2018/05/25/3-things-two-men-learned-from-their-650000-lunch-with-warren-buffett.html

Spier, G. (2018, June 12). *Opinion: What lunch with Warren Buffett taught me about investing and life.* MarketWatch. https://www.marketwatch.com/story/what-lunch-with-warren-buffett-taught-me-about-investing-and-life-2018-06-11

Mejia, Z. (2018, May 29). *Warren Buffett has raised $26 million auctioning lunch dates—here's how the tradition started.* Make It. CNBC.com. https://www.cnbc.com/2018/05/29/why-warren-buffett-auctions-million-dollar-lunch-dates-for-charity.html

Stempel, J. (2019, June 1). *Warren Buffett charity lunch fetches record $4.57 million winning bid.* Reuters. https://www.reuters.com/article/us-buffett-lunch/warren-buffett-charity-lunch-fetches-record-4-57-million-winning-bid-idUSKCN1T238Y

Chapter 5

Physicists are designing experiments to turn protons into electrons: Luo, W., Wu, S. D., Liu, W. Y., Ma, Y. Y., Li, F. Y., Yuan, T., . . . Sheng, Z. M. (2018). Enhanced electron-positron pair production by two obliquely incident lasers interacting with a solid target. *Plasma Physics and Controlled Fusion, 60*(9), 095006.

Hormonal effects can extend from a few seconds to a few days: Singh, G. K. (2016). What is the difference between hormones and neurotransmitters? Retrieved Jan 2, 2020 from https://www.quora.com/What-is-the-difference-between-hormones-and-neurotransmitter

Neurotransmitters travel only microscopic distances: Purves, D., Augustine, G. J., Fitzpatrick, D., Hall, W. C., Lamantia, A. S., Mooney, R. D., Platt, M. L., & White, L. E. (Eds.). (2001). What defines a neurotransmitter? Neuroscience (2nd ed.). Sunderland, MA: Sinauer Associates. https://www.ncbi.nlm.nih.gov/books/NBK10957

Some molecules operate as both hormones and neurotransmitters: Singh, G. K. (2016). What is the difference between hormones and neurotransmitters? Retrieved Jan 2, 2020 from https://www.quora.com/What-is-the-difference-between-hormones-and-neurotransmitter

Glutamate is the most excitatory neurotransmitter and GABA is the most inhibitory: Sapolsky, R. M. (2018). *Behave: The biology of humans at our best and worst.* New York, NY: Penguin.

GABA is the "brain's own Valium": Blesching, U. (2015). *The cannabis health index: Combining the science of medical marijuana with mindfulness techniques to heal 100 chronic symptoms and diseases.* Novato, CA: New World Library.

Time it takes for neurotransmitters to act: Sapolsky, R. M. (2018). *Behave: The biology of humans at our best and worst.* New York, NY: Penguin.

Twarog's discovery was "the cornerstone of the antidepressant revolution": Greenberg, G. (2011). *Manufacturing depression: The secret history of a modern disease.* New York, NY: Simon and Schuster.

Pavlov's Nobel speech: Pavlov, I. (1904). *Physiology of digestion.* The Nobel Prize. Retrieved Jan 2, 2020 from https://www.nobelprize.org/prizes/medicine/1904/pavlov/lecture

Serotonin identified in the gastrointestinal tract in 1935: Beck, T. (2015, December 17). *A vaccine for depression? Ketamine's remarkable effect bolsters a new theory of mental illness. Nautilus, 31.* http://nautil.us//issue/31/stress/a-vaccine-for-depression

Twarog identified serotonin in the brains of monkeys, rats, and dogs: Greenberg, G. (2011). *Manufacturing depression: The secret history of a modern disease.* New York, NY: Simon and Schuster.

The journal editor refused to have reviewers look at a paper by a girl: Greenberg, G. (2011). *Manufacturing depression: The secret history of a modern disease.* New York, NY: Simon and Schuster.

Sales and effects of psychotropic drugs: Whitaker, R. (2011). *Anatomy of an epidemic: Magic bullets, psychiatric drugs, and the astonishing rise of mental illness in America.* New York, NY: Random House.

The erroneous notion that chemical imbalance in the brain causes mental disorders: Pies, R. W. (2011, July 11). Psychiatry's new brain-mind and the legend of the "chemical imbalance." *Psychiatric Times.* https://www.psychiatrictimes.com/articles/psychiatrys-new-brain-mind-and-legend-chemical-imbalance

The damaging and far-reaching effects of the characterization of mental disorders as disease: Maisel, E. (2016, March 8). Gary Greenberg on manufacturing depression. *Psychology Today.* https://www.psychologytoday.com/us/blog/rethinking-mental-health/201603/gary-greenberg-manufacturing-depression

DHEA's antiaging effects: Forti, P., Maltoni, B., Olivelli, V., Pirazzoli, G. L., Ravaglia, G., & Zoli, M. (2012). Serum dehydroepiandrosterone sulfate and adverse health outcomes in older men and women. *Rejuvenation Research, 15*(4), 349–358.

DHEA deficiency has been linked to cancer, heart disease, obesity, and diabetes: Rizvi, S. I., & Jha, R. (2011). Strategies for the discovery of anti-aging compounds. *Expert Opinion on Drug Discovery, 6*(1), 89–102.

Do Vale, S., Martin, J. M., Fagundes, M. J., & do Carmo, I. (2011). Plasma dehydroepiandrosterone-sulphate is related to personality and stress response. *Neuroendocrinology Letters, 32*(4), 442–448.

Exogenous molecules like opioids are effective because they fit into the same receptor sites: Sapolsky, R. M. (2018). *Behave: The biology of humans at our best and worst.* New York, NY: Penguin.

The effects of prescription and recreational drugs on excitatory and inhibitory neurotransmitters: Sherman, C. (2017, March 9). *Impacts of drugs on neurotransmission.*

National Institute on Drug Abuse. Retrieved Jan 2, 2020 from https://www.drugabuse.gov/news-events/nida-notes/2017/03/impacts-drugs-neurotransmission

The effect of alcohol on the various neurotransmitters: AccessMedicine. (2018). *The effect of alcohol on neurotransmitters in the brain.* Retrieved Jan 2, 2020 from https://www.accessmedicinenetwork.com/users/82976-harrison-s-self-assessment-and-board-review/posts/34085-the-effect-of-alcohol-on-neurotransmitters-in-the-brain

Types of benzodiazepines: Tuck. (2017). Benzodiazepines—types, side effects, addiction, and withdrawal. Retrieved Jan 7, 2020 from https://www.tuck.com/benzodiazepines

The chemical similarities between LSD and serotonin: Beck, T. (2015, December 17). A vaccine for depression? Ketamine's remarkable effect bolsters a new theory of mental illness. *Nautilus, 31.* Retrieved Jan 4, 2020 from http://nautil.us//issue/31/stress/a-vaccine-for-depression

As with neurotransmitters, the ratios between hormones are more important: Kamin, H. S., & Kertes, D. A. (2017). Cortisol and DHEA in development and psychopathology. *Hormones and Behavior, 89,* 69–85.

Meditation has been shown to produce a rise in norepinephrine in the brain: Hölzel, B. K., Carmody, J., Vangel, M., Congleton, C., Yerramsetti, S. M., Gard, T., & Lazar, S. W. (2011). Mindfulness practice leads to increases in regional brain gray matter density. *Psychiatry Research: Neuroimaging, 191*(1), 36-43.

The alpha bridge: Cade, M., & Coxhead, N. (1979). *The awakened mind: Biofeedback and the development of higher states of awareness.* New York, NY: Dell.

Brain waves associated with hormone and neurotransmitter changes: Sakuta, A. (2016). Neurophenomenology of flow and meditative moving [lecture]. University of Chichester. Retrieved April 8, 2019, from eprints.chi.ac.uk

Areas *of the brain are integral to the dopamine reward system:* Newberg, A., & Waldman, M. R. (2017). *How enlightenment changes your brain: The new science of transformation.* New York, NY: Penguin.

Anticipation of financial reward can activate dopamine: Knutson, B., Adams, C. M., Fong, G. W., & Hommer, D. (2001). Anticipation of increasing monetary reward selectively recruits nucleus accumbens. *Journal of Neuroscience, 21*(16), RC159.

Dopamine levels can be depleted by chronic pain or chronic stress: Sapolsky, R. M. (2018). *Behave: The biology of humans at our best and worst.* New York, NY: Penguin.

Dopamine is about the happiness of pursuit of reward: Sapolsky, R. M. (2018). *Behave: The biology of humans at our best and worst.* New York, NY: Penguin.

Serotonin and behavior: Bortolato, M., Pivac, N., Seler, D. M., Perkovic, M. N., Pessia, M., & Di Giovanni, G. (2013). The role of the serotonergic system at the interface of aggression and suicide. *Neuroscience, 236,* 160–185.

Low serotonin is linked to aggression and suicide:

Bortolato, M., Pivac, N., Seler, D. M., Perkovic, M. N., Pessia, M., & Di Giovanni, G. (2013). The role of the serotonergic system at the interface of aggression and suicide. *Neuroscience, 236,* 160–185.

Åsberg, M., Träskman, L., & Thorén, P. (1976). 5-HIAA in the cerebrospinal fluid: A biochemical suicide predictor? *Archives of General Psychiatry, 33*(10), 1193–1197.

Gordon Wasson's experiences with magic mushrooms: Wasson, R. G. (2015, April 4). Seeking the magic mushroom. *Door of Perception.* Retrieved Jan 7, 2020 from https://doorofperception.com/2015/04/r-gordon-wasson-seeking-the-magic-mushroom

The role of norepinephrine in amplifying focus: Kotler, S., & Wheal, J. (2017). *Stealing fire: How Silicon Valley, the Navy SEALs, and maverick scientists are revolutionizing the way we live and work.* New York, NY: HarperCollins.

The brains of meditators produce norepinephrine: Lazar, S. (2014). Change in brainstem gray matter concentration following a mindfulness-based intervention is correlated with improvement in psychological well-being. *Frontiers in Human Neuroscience, 8,* 33.

Oxytocin and behavior: Sapolsky, R. M. (2018). *Behave: The biology of humans at our best and worst.* New York, NY: Penguin.

Couples in love and oxytocin: Esch, T., & Stefano, G. B. (2005). The neurobiology of love. *Neuroendocrinology Letters, 26*(3), 175–192.

Oxytocin and erections: Filippi, S., Vignozzi, L., Vannelli, G. B., Ledda, F., Forti, G., & Maggi, M. (2003). Role of oxytocin in the ejaculatory process. *Journal of Endocrinological Investigation, 26*(3 Suppl), 82–86.

Tapping the cheek, as we do in EFT, produced an 800% spike in delta: Harper, M. (2012). Taming the amygdala: An EEG analysis of exposure therapy for the traumatized. *Traumatology, 18*(2), 61–74.

The orgasmic brain: Mitrokostas, S. (2019, January 24). *12 things that happen in your brain when you have an orgasm.* Business Insider. Retrieved Jan 12, 2020 from https://www.businessinsider.com/what-happens-to-your-brain-during-orgasm-2019-1#when-you-orgasm-your-brain-releases-a-surge-of-dopamine-3

Oxytocin clears glutamate: Theodosis, D. T. (2002). Oxytocin-secreting neurons: A physiological model of morphological neuronal and glial plasticity in the adult hypothalamus. *Frontiers in Neuroendocrinology, 23*(1), 101–135.

Oxytocin stimulates the release of nitric oxide in blood vessels: Thibonnier, M., Conarty, D. M., Preston, J. A., Plesnicher, C. L., Dweik, R. A., & Erzurum, S. C. (1999). Human vascular endothelial cells express oxytocin receptors. *Endocrinology, 140*(3), 1301–1309.

A study of 34 married couples and "listening touch": Holt-Lunstad, J., Birmingham, W. A., & Light, K. C. (2008). Influence of a "warm touch" support enhancement intervention among married couples on ambulatory blood pressure, oxytocin, alpha amylase, and cortisol. *Psychosomatic Medicine, 70*(9), 976–985.

Human and dog gazers: Nagasawa, M., Mitsui, S., En, S., Ohtani, N., Ohta, M., Sakuma, Y., . . . Kikusui, T. (2015). Oxytocin-gaze positive loop and the coevolution of human-dog bonds. *Science, 348*(6232), 333–336.

Oxytocin spikes anandamide: Wei, D., Lee, D., Cox, C. D., Karsten, C. A., Peñagarikano, O., Geschwind, D. H., . . . Piomelli, D. (2015). Endocannabinoid signaling mediates oxytocin-driven social reward. *Proceedings of the National Academy of Sciences, 112*(45), 14084–14089.

Beta-endorphin is 18 to 33 times more powerful than morphine: Loh, H. H., Tseng, L. F., Wei, E., & Li, C. H. (1976). Beta-endorphin is a potent analgesic agent. *Proceedings of the National Academy of Sciences, 73*(8), 2895–2898.

Beta-endorphin is 48 times more powerful than morphine: Loh, H. H., Brase, D. A., Sampath-Khanna, S., Mar, J. B., Way, E. L., & Li, C. H. (1976). β-Endorphin in vitro inhibition of striatal dopamine release. *Nature, 264*(5586), 567.

Anandamide is involved in the neural generation of pleasure and motivation: Mukerji, S. (n.d.). Anandamide–the joy chemical. Retrieved Jan 7, 2020 from http://www.sohamhappinessprogram.com/anandamide-joy-chemical

Nitric oxide, neuroplasticity, and oxygen flow: Petrie, M., Rejeski, W. J., Basu, S., Laurienti, P. J., Marsh, A. P., Norris, J. L., . . . Burdette, J. H. (2016). Beet root juice: An ergogenic aid for exercise and the aging brain. *Journals of Gerontology Series A: Biomedical Sciences and Medical Sciences, 72*(9), 1284–1289.

Health benefits of nitric oxide:

Coleman, J. W. (2001). Nitric oxide in immunity and inflammation. *International Immunopharmacology, 1*(8), 1397–1406. doi:10.1016/S1567-5769(01)00086-8.

Alam, M. S., Akaike, T., Okamoto, S., Kubota, T., Yoshitake, J., Sawa, T., . . . Maeda, H. (2002). Role of nitric oxide in host defense in murine salmonellosis as a function of its antibacterial and antiapoptotic activities. *Infection and Immunity, 70*(6), 3130–3142.

Bryan, N. S. (2006). Nitrite in nitric oxide biology: Cause or consequence?: A systems-based review. *Free Radical Biology and Medicine, 41*(5), 691–701.

Webb, A. J., Patel, N., Loukogeorgakis, S., Okorie, M., Aboud, Z., Misra, S., . . . MacAllister, R. (2008). Acute blood pressure lowering, vasoprotective, and antiplatelet properties of dietary nitrate via bioconversion to nitrite. *Hypertension, 51*(3), 784–790.

Drop-off of nitric oxide as we age: Mercola, J. (2019, January 14). *Top 9 reasons to optimize your nitric oxide production.* Retrieved Jan 2, 2020 from https://articles.mercola.com/sites/articles/archive/2019/01/14/get-nourished-with-nitrates.aspx

Benefits of plant-based nitric oxide: Mercola, J. (2019, January 14). *Top 9 reasons to optimize your nitric oxide production.* Retrieved Jan 6, 2020 from https://articles.mercola.com/sites/articles/archive/2019/01/14/get-nourished-with-nitrates.aspx

Associations between brain waves and neurotransmitters: Sakuta, A. (2016). Neurophenomenology of flow and meditative Moving [lecture]. University of Chichester. Retrieved Jan 3, 2020 from eprints.chi.ac.uk

The role of the endocannabinoid system in appetite, pain, inflammation, sleep, stress, mood, memory, motivation, and reward: Fallis, J. (2017, July 20). 25 powerful ways to boost your endocannabinoid system. *Optimal Living Dynamics.* Retrieved Jan 7, 2020 from https://www.optimallivingdynamics.com/blog/how-to-stimulate-and-support-your-endocannabinoid-system

Locations of CB1 and CB2 receptors: Hanuš, L. O. (2007). Discovery and isolation of anandamide and other endocannabinoids. *Chemistry and Biodiversity, 4*(8), 18281–841.

Low endocannabinoid levels are linked to major depression, generalized anxiety disorder, PTSD, multiple sclerosis, attention deficit/hyperactivity disorder (ADHD), Parkinson's disease, fibromyalgia, and sleep disorders: Fallis, J. (2017, July 20). 25 powerful ways to boost your endocannabinoid system. *Optimal Living Dynamics.* Retrieved Jan 7, 2020 from https://www.optimallivingdynamics.com/blog/how-to-stimulate-and-support-your-endocannabinoid-system

Foods that stimulate the endocannabinoid system: Fallis, J. (2017, July 20). 25 powerful ways to boost your endocannabinoid system. *Optimal Living Dynamics.* Retrieved Jan 5, 2020 from https://www.optimallivingdynamics.com/blog/how-to-stimulate-and-support-your-endocannabinoid-system

Behavioral methods that stimulate the endocannabinoid system: Mukerji, S. (n.d.). Anandamide–the joy chemical [Web log post]. Retrieved from http://www.sohamhappinessprogram.com/anandamide-joy-chemical

Oxytocin-driven anandamide signaling: Wei, D., Lee, D., Cox, C. D., Karsten, C. A., Peñagarikano, O., Geschwind, D. H., . . . Piomelli, D. (2015). Endocannabinoid signaling mediates oxytocin-driven social reward. *Proceedings of the National Academy of Sciences, 112*(45), 14084–14089.

Lumír Hanuš and the bliss molecule: D'Mikos, S. (2019). *The discovery of anandamide* [video]. Vimeo. Retrieved Jan 7, 2020 from https://vimeo.com/328088031

The discovery of anandamide in the human brain: Hurt, L. (2016, December 13). Meet Lumir Hanus, who discovered the first endocannabinoid. *Leafly.* Retrieved Jan 2, 2020 from https://www.leafly.com/news/science-tech/lumir-hanus-discovered-first-endocannabinoid-anandamide

Meditation and elevated dopamine levels: Kjaer, T. W., Bertelsen, C., Piccini, P., Brooks, D., Alving, J., & Lou, H. C. (2002). Increased dopamine tone during meditation-induced change of consciousness. *Cognitive Brain Research, 13*(2), 255–259.

Review and synthesis of the research literature on neural and hormonal changes associated with meditation: Newberg, A. B., & Iversen, J. (2003). The neural basis of the complex mental task of meditation: neurotransmitter and neurochemical considerations. *Medical Hypotheses, 61*(2), 282–291.

Meditation increases nitric oxide: Chiesa, A., & Serretti, A. (2010). A systematic review of neurobiological and clinical features of mindfulness meditations. *Psychological Medicine, 40*(8), 1239–1252.

Oxytocin is increased by meditation, and triggers the release of other pleasure chemicals: Millière, R., Carhart-Harris, R. L., Roseman, L., Trautwein, F. M., & Berkovich-Ohana, A. (2018). Psychedelics, meditation, and self-consciousness. *Frontiers in Psychology, 9.*

Ashar, Y. K., Andrews-Hanna, J. R., Yarkoni, T., Sills, J., Halifax, J., Dimidjian, S., & Wager, T. D. (2016). Effects of compassion meditation on a psychological model of charitable donation. *Emotion, 16*(5), 691.

Van Cappellen, P., Way, B. M., Isgett, S. F., & Fredrickson, B. L. (2016). Effects of oxytocin administration on spirituality and emotional responses to meditation. *Social Cognitive and Affective Neuroscience, 11*(10), 1579–1587.

Nitric oxide release is closely coupled with anandamide production; thus meditation and other stress-reducing activities may stimulate both: Esch, T., & Stefano, G. B. (2010). The neurobiology of stress management. *Neuroendocrinology Letters, 31*(1), 19–39.

Heightened oxytocin mobilizes the synthesis of anandamide: Wei, D., Lee, D., Cox, C. D., Karsten, C. A., Peñagarikano, O., Geschwind, D. H., . . . Piomelli, D. (2015). Endocannabinoid signaling mediates oxytocin-driven social reward. *Proceedings of the National Academy of Sciences, 112*(45), 14084–14089.

Anandamide can also improve cognitive function, motivation, learning, and memory: Mechoulam, R., & Parker, L. A. (2013). The endocannabinoid system and the brain. *Annual Review of Psychology, 64,* 21–47.

Meditation stimulates the brain to produce many rewarding chemicals at the same time: Ospina, M. B., Bond, K., Karkhaneh, M., Tjosvold, L., Vandermeer, B., Liang, Y., . . . Klassen, T. P. (2007). Meditation practices for health: State of the research. *Evidence Report/Technology Assessment,* 155-163.

Jindal, V., Gupta, S., & Das, R. (2013). Molecular mechanisms of meditation. *Molecular Neurobiology, 48*(3), 808–811.

Researchers using fMRIs find that bathing the brain in the chemicals of bliss conditions its functioning the rest of the day: Britton, W. B., Lindahl, J. R., Cahn, B. R., Davis, J. H., & Goldman, R. E. (2014). Awakening is not a metaphor: the effects of Buddhist meditation practices on basic wakefulness. *Annals of the New York Academy of Sciences, 1307*(1), 64–81.

Harvard professor Teresa Amabile found that people in flow were still more creative the following day: Amabile, T. M., & Pillemer, J. (2012). Perspectives on the social psychology of creativity. *Journal of Creative Behavior, 46*(1), 3–15.

The meditative state is akin to the state of "flow" or being in "the zone": Edinger-Schons, L. M. (2019). Oneness beliefs and their effect on life satisfaction. *Psychology of Religion and Spirituality 11*(1). DOI: 10.1037/rel0000259

Columbia University bioengineering lab virtual reality flight study comments: Columbia University School of Engineering and Applied Science. (2019, March 12). *Neurofeedback gets you back in the zone: New study from biomedical engineers demonstrates that a brain-computer interface can improve your performance.* ScienceDaily. Retrieved May 13, 2019, from www.sciencedaily.com/releases/2019/03/190312143206.htm

McKinsey study of high-performance executives called upon to solve difficult strategic problems: Cranston, S., & Keller, S. (2013). Increasing the meaning quotient of work. *McKinsey Quarterly, 1*(48–59).

Students solved a conceptual problem eight times better: Chi, R. P., & Snyder, A. W. (2012). Brain stimulation enables the solution of an inherently difficult problem. *Neuroscience Letters, 515*(2), 121–124.

DARPA neurofeedback study of complex problem solving in flow states: Adee, S. (2012). Zap your brain into the zone: Fast track to pure focus. *New Scientist, 2850,* 1–6.

LSD, psilocybin, and DMT activate serotonin receptors: Carhart-Harris, R. L. (2018). Serotonin, psychedelics and psychiatry. *World Psychiatry: Official Journal of the World Psychiatric Association (WPA), 17*(3), 358–359.

LSD, psilocybin, and DMT activate norepinephrine receptors, while the cannabinoids in marijuana use anandamide receptors: King, C. (2013). The cosmology of conscious mental states (Part I). *Journal of Consciousness Exploration and Research, 4*(6), 561–581.

The two plants combined to form ayahuasca and their molecular properties: Riba, J., Valle, M., Urbano, G., Yritia, M., Morte, A., & Barbanoj, M. J. (2003). Human pharmacology of ayahuasca: Subjective and cardiovascular effects, monoamine metabolite excretion, and pharmacokinetics. *Journal of Pharmacology and Experimental Therapeutics, 306*(1), 73–83.

Physiologically, the path to bliss is the same whether brought on by drugs, meditation, or another mind-altering experience: King, C. (2013). The cosmology of conscious mental states (Part I). *Journal of Consciousness Exploration and Research, 4*(6), 561–581.

Hamlet on hash: Thackeray, F. (2015). Shakespeare, plants, and chemical analysis of early 17th century clay 'tobacco' pipes from Europe. *South African Journal of Science, 111*(7/8), article a0115. doi:10.17159/sajs.2015/a0115

Thackeray, J. F., Van der Merwe, N. J., & Van der Merwe, T. A. (2001). Chemical analysis of residues from seventeenth century clay pipes from Stratford-upon-Avon and environs. *South African Journal of Science*, 97, 19–21.

Time *magazine article on Shakespeare's pipes:* Jenkins, N. (2015, August 10). Scientists detect traces of cannabis on pipes found in William Shakespeare's garden. *Time.* http://time.com/3990305/william-shakespeare-cannabis-marijuana-high

Just a couple of joints of marijuana can affect the adolescent brain: Orr, C., Spechler, P., Cao, Z., Albaugh, M., Chaarani, B., Mackey, S., . . . Bromberg, U. (2019). Grey matter volume differences associated with extremely low levels of cannabis use in adolescence. *Journal of Neuroscience, 39*(10), 1817–1827.

Teenagers using cannabis had a nearly 40% greater risk of depression and a 50% greater risk of suicidal ideation: Gobbi, G., Atkin, T., Zytynski, T., Wang, S., Askari, S., Boruff, J., . . . Mayo, N. (2019, February 13). Association of cannabis use in adolescence and risk of depression, anxiety, and suicidality in young adulthood: A systematic review and meta-analysis. *JAMA Psychiatry.* doi:10.1001/jamapsychiatry.2018.4500

With prolonged marijuana use in adults, the wiring of the brain degrades: Filbey, F. M., Aslan, S., Calhoun, V. D., Spence, J. S., Damaraju, E., Caprihan, A., & Segall, J. (2014). Long-term effects of marijuana use on the brain. *Proceedings of the National Academy of Sciences, 111*(47), 16913–16918.

The authors of one study found that regular cannabis use is associated with gray matter volume reduction: Battistella, G., Fornari, E., Annoni, J. M., Chtioui, H., Dao, K., Fabritius, M., . . . Giroud, C. (2014). Long-term effects of cannabis on brain structure. *Neuropsychopharmacology, 39*(9), 2041.

Cannabis use both increases anxiety and depression and leads to worse health: Black, N., Stockings, E., Campbell, G., Tran, L. T., Zagic, D., Hall, W. D., . . . Degenhardt, L. (2019). Cannabinoids for the treatment of mental disorders and symptoms of mental disorders: a systematic review and meta-analysis. *The Lancet Psychiatry, 6*(12), 995-1010.

Research into MDMA has shown that it can produce serious side effects: Parrott, A. C. (2013). Human psychobiology of MDMA or "Ecstasy": An overview of 25 years of empirical research. *Human Psychopharmacology: Clinical and Experimental, 28*(4), 289–307.

Chapter 6

Marian Diamond: Diamond, M. C. (n.d.). Marian Cleeves Diamond [autobiographical essay]. Retrieved Jan 7, 2020 from https://www.sfn.org/~/media/SfN/Documents/TheHistoryofNeuroscience/Volume%206/c3.ashx

Grimes, W. (2017, August 16). Marian C. Diamond, 90, student of the brain, is dead. *The New York Times*. Retrieved Jan 7, 2020 from https://www.nytimes.com/2017/08/16/science/marian-c-diamond-90-student-of-the-brain-is-dead.html

Sanders, R. (2017, July 28). Marian Diamond, known for studies of Einstein's brain, dies at 90. *Berkeley News*. Retrieved Jan 2, 2020 from https://news.berkeley.edu/2017/07/28/marian-diamond-known-for-studies-of-einsteins-brain-dies-at-90

Scheidt, R. J. (2015). Marian Diamond: The "Mitochondrial Eve" of successful aging. *Gerontologist, 55*(1). doi:10.1093/geront/gnu124

Translational gap: Church, D., Feinstein, D., Palmer-Hoffman, J., Stein, P. K., & Tranguch, A. (2014). Empirically supported psychological treatments: The challenge of evaluating clinical innovations. *Journal of Nervous and Mental Disease, 202*(10), 699–709. doi:10.1097/NMD.0000000000000188

State Progression: Hall, M. L. (2009). *Achieving peak performance: The science and art of taking performance to ever higher levels.* Bethel, CT: Crown House.

Altered Traits: Haidt, J. (2012). *The righteous mind: Why good people are divided by politics and religion.* New York, NY: Vintage.

Schwartz, J. M., Stapp, H. P., & Beauregard, M. (2005). Quantum physics in neuroscience and psychology: A neurophysical model of mind-brain interaction. *Philosophical Transactions of the Royal Society of London B: Biological Sciences, 360*(1458), 1309–1327.

Sara Lazar: Lazar, S. W., Kerr, C. E., Wasserman, R. H., Gray, J. R., Greve, D. N., Treadway, M. T., . . . Rauch, S. L. (2005). Meditation experience is associated with increased cortical thickness. *Neuroreport, 16*(17), 1893–1897.

Hölzel, B. K., Carmody, J., Vangel, M., Congleton, C., Yerramsetti, S. M., Gard, T., & Lazar, S. W. (2011). Mindfulness practice leads to increases in regional brain gray matter density. *Psychiatry Research: Neuroimaging, 191*(1), 36–43.

Schulte, B. (2015, May 26). Harvard neuroscientist: Meditation not only reduces stress, here's how it changes your brain. *Washington Post.* Retrieved Jan 9, 2020 from https://www.washingtonpost.com/news/inspired-life/wp/2015/05/26/harvard-neuroscientist-meditation-not-only-reduces-stress-it-literally-changes-your-brain

Cromie, W. J. (2006, February 2). Meditation found to increase brain size. *Harvard Gazette.* Retrieved Jan 9, 2020 from https://news.harvard.edu/gazette/story/2006/02/meditation-found-to-increase-brain-size

A meta-analysis of brain imaging studies: Boccia, M., Piccardi, L., & Guariglia, P. (2015). The meditative mind: A comprehensive meta-analysis of MRI studies. *BioMed Research International, 2015,* 419808. doi:10.1155/2015/419808

Kirtan Kriya study: Khalsa, D. S., Amen, D., Hanks, C., Money, N., & Newberg, A. (2009). Cerebral blood flow changes during chanting meditation. *Nuclear Medicine Communications, 30*(12), 956-961.

Size Does Matter: Men, W., Falk, D., Sun, T., Chen, W., Li, J., Yin, D., . . . Fan, M. (2014). The corpus callosum of Albert Einstein's brain: Another clue to his high intelligence? *Brain, 137*(4), e268-e278.

Falk, D., Lepore, F. E., & Noe, A. (2012). The cerebral cortex of Albert Einstein: A description and preliminary analysis of unpublished photographs. *Brain, 136*(4), 1304–1327. https://www.ncbi.nlm.nih.gov/pmc/articles/PMC3613708

Witelson, S. F., Kigar, D. L., & Harvey, T. (1999). The exceptional brain of Albert Einstein. *Lancet, 353*(9170), 2149–2153.

Healy, M. (2013, October 11). Einstein's brain really was bigger than most people's. *Seattle Times*. Retrieved Jan 9, 2020 from https://www.seattletimes.com/nation-world/einsteinrsquos-brain-really-was-bigger-than-most-peoplersquos

The Four Key Networks: Goleman, D., & Davidson, R. J. (2018). *Altered traits: Science reveals how meditation changes your mind, brain, and body.* New York, NY: Avery.

Newberg, A., & Waldman, M. R. (2017). *How enlightenment changes your brain: The new science of transformation.* New York, NY: Penguin.

Emotion Regulation Network References

Hostile takeover of consciousness by emotion: LeDoux, J. E. (2002). *Synaptic self: How our brains become who we are.* New York, NY: Penguin.

Hippocampus shrinks with depression: Sheline, Y. I., Liston, C., & McEwen, B. S. (2019). Parsing the hippocampus in depression: Chronic stress, hippocampal volume, and major depressive disorder. *Biological Psychiatry, 85*(6), 436–438.

Meditation leads to increases in gray matter in hippocampus: Hölzel, B. K., Carmody, J., Vangel, M., Congleton, C., Yerramsetti, S. M., Gard, T., & Lazar, S. W. (2011). Mindfulness practice leads to increases in regional brain gray matter density. *Psychiatry Research: Neuroimaging, 191*(1), 36–43.

Pickut, B. A., Van Hecke, W., Kerckhofs, E., Mariën, P., Vanneste, S., Cras, P., & Parizel, P. M. (2013). Mindfulness based intervention in Parkinson's disease leads to structural brain changes on MRI: A randomized controlled longitudinal trial. *Clinical Neurology and Neurosurgery, 115*(12), 2419–2425. doi:10.1016/j.clineuro.2013.10.002

Thalamus: Newberg, A., & Waldman, M. R. (2017). *How enlightenment changes your brain: The new science of transformation.* New York, NY: Penguin.

Amygdala connections: Goleman, D., & Davidson, R. J. (2018). *Altered traits: Science reveals how meditation changes your mind, brain, and body.* New York, NY: Avery.

The true mark of a meditator is that he has disciplined his mind: Goleman, D., & Davidson, R. J. (2018). *Altered traits: Science reveals how meditation changes your mind, brain, and body.* New York, NY: Avery.

Amygdala can also be regulated by the striatum, especially the basal ganglia: Rodriguez-Romaguera, J., Do Monte, F. H., & Quirk, G. J. (2012). Deep brain stimulation of the ventral striatum enhances extinction of conditioned fear. *Proceedings of the National Academy of Sciences, 109*(22), 8764-8769.

One eighth: St. Abba Dorotheus, quoted in Kadloubovsky, E., & Palmer, G. E. H. (1971). *Early fathers from the Philokalia.* London, UK: Faber & Faber.

Graham Phillips: Catalyst. (2016, June 7). *The science of meditation – can it really change you?* Retrieved Jan 2, 2020 from http://www.abc.net.au/catalyst/stories/4477405.htm

Dentate gyrus image: Noguchi, H., Murao, N., Kimura, A., Matsuda, T., Namihira, M., & Nakashima, K. (2016). DNA methyltransferase 1 is indispensable for development of the hippocampal dentate gyrus. *Journal of Neuroscience, 36*(22), 6050-6068.

Brain regions that grow:

Ventromedial prefrontal cortex: Chau, B. K., Keuper, K., Lo, M., So, K. F., Chan, C. C., & Lee, T. M. (2018). Meditation-induced neuroplastic changes of the prefrontal network are associated with reduced valence perception in older people. *Brain and Neuroscience Advances, 2.* doi:10.1177/2398212818771822

Thalamus: Luders, E., Toga, A. W., Lepore, N., & Gaser, C. (2009). The underlying anatomical correlates of long-term meditation: Larger hippocampal and frontal volumes of gray matter. *Neuroimage, 45*(3), 672–678.

Temporoparietal junction: Hölzel, B. K., Carmody, J., Vangel, M., Congleton, C., Yerramsetti, S. M., Gard, T., & Lazar, S. W. (2011). Mindfulness practice leads to increases in regional brain gray matter density. *Psychiatry Research: Neuroimaging, 191*(1), 36–43.

Boccia, M., Piccardi, L., & Guariglia, P. (2015). The meditative mind: A comprehensive meta-analysis of MRI studies. *BioMed Research International, 2015,* 419808. doi:10.1155/2015/419808

Hippocampus: Fox, K. C., Nijeboer, S., Dixon, M. L., Floman, J. L., Ellamil, M., Rumak, S. P., Sedlmeier, P., & Christoff, K. (2014). Is meditation associated with altered brain structure? A systematic review and meta-analysis of morphometric neuroimaging in meditation practitioners. *Neuroscience and Biobehavioral Reviews, 43,* 48–73. doi:10.1016/j.neubiorev.2014.03.016

Subiculum: Boccia, M., Piccardi, L., & Guariglia, P. (2015). The meditative mind: A comprehensive meta-analysis of MRI studies. *BioMed Research International, 2015,* 419808. doi:10.1155/2015/419808.

Anterior and mid cingulate: Fox, K. C., Nijeboer, S., Dixon, M. L., Floman, J. L., Ellamil, M., Rumak, S. P., Sedlmeier, P., & Christoff, K. (2014). Is meditation associated with altered brain structure? A systematic review and meta-analysis of morphometric neuroimaging in meditation practitioners. *Neuroscience and Biobehavioral Reviews, 43,* 48–73. doi:10.1016/j.neubiorev.2014.03.016

Orbitofrontal cortex: Fox et al., 2014; Luders, Toga, Lepore, & Gaser, 2009.

Precuneus: Kurth, F., Luders, E., Wu, B., & Black, D. S. (2014). Brain gray matter changes associated with mindfulness meditation in older adults: An exploratory pilot study using voxel-based morphometry. *Neuro: Open journal, 1*(1), 23–26. doi:10.17140/NOJ-1-106

Ventromedial orbitofrontal cortex: Hernández, S. E., Suero, J., Barros, A., González-Mora, J. L., & Rubia, K. (2016). Increased grey matter associated with long-term sahaja yoga meditation: A voxel-based morphometry study. *PloS One, 11*(3), e0150757. doi:10.1371/journal.pone.0150757

The connections between the prefrontal cortex and the amygdala strengthen: Gotink, R. A., Meijboom, R., Vernooij, M. W., Smits, M., & Hunink, M. M. (2016). 8-week mindfulness based stress reduction induces brain changes similar to traditional long-term meditation practice–a systematic review. *Brain and Cognition, 108,* 32–41. doi:10.1016/j.bandc.2016.07.001

The stronger this link is, the less reactive you become: Goleman & Davidson, 2018, p. 97.

Attention Network References

Brain regions that grow:

Insula: Gotink, R. A., Meijboom, R., Vernooij, M. W., Smits, M., & Hunink, M. M. (2016). 8-week mindfulness based stress reduction induces brain changes similar to traditional long-term meditation practice–a systematic review. *Brain and Cognition, 108,* 32–41. doi:10.1016/j.bandc.2016.07.001

Cortical somatomotor areas: Fox, K. C., Nijeboer, S., Dixon, M. L., Floman, J. L., Ellamil, M., Rumak, S. P., Sedlmeier, P., & Christoff, K. (2014). Is meditation associated with altered brain structure? A systematic review and meta-analysis of morphometric neuroimaging in meditation practitioners. *Neuroscience and Biobehavioral Reviews, 43,* 48–73. doi:10.1016/j.neubiorev.2014.03.016

Prefrontal cortex centerline regions: Tang, Y. Y., Hölzel, B. K., & Posner, M. I. (2015). The neuroscience of mindfulness meditation. *Nature Reviews Neuroscience, 16*(4), 213.

Posterior cingulate cortex: Hölzel, B. K., Carmody, J., Vangel, M., Congleton, C., Yerramsetti, S. M., Gard, T., & Lazar, S. W. (2011). Mindfulness practice leads to increases in regional brain gray matter density. *Psychiatry Research: Neuroimaging, 191*(1), 36–43. Tang et al., 2015.

Precuneus: Kurth, F., Luders, E., Wu, B., & Black, D. S. (2014). Brain gray matter changes associated with mindfulness meditation in older adults: An exploratory pilot study using voxel-based morphometry. *Neuro: Open journal, 1*(1), 23–26. doi:10.17140/NOJ-1-106

Anterior and mid cingulate and orbitofrontal cortex: Fox, K. C., Nijeboer, S., Dixon, M. L., Floman, J. L., Ellamil, M., Rumak, S. P., Sedlmeier, P., & Christoff, K. (2014). Is meditation associated with altered brain structure? A systematic review and meta-analysis of morphometric neuroimaging in meditation practitioners. *Neuroscience and Biobehavioral Reviews, 43,* 48–73. doi:10.1016/j.neubiorev.2014.03.016

Angular gyrus: Boccia, M., Piccardi, L., & Guariglia, P. (2015). The meditative mind: A comprehensive meta-analysis of MRI studies. *BioMed Research International, 2015,* 419808. doi:10.1155/2015/419808

Pons: Singleton, O., Hölzel, B. K., Vangel, M., Brach, N., Carmody, J., & Lazar, S. W. (2014). Change in brainstem gray matter concentration following a mindfulness-based intervention is correlated with improvement in psychological well-being. *Frontiers in Human Neuroscience, 8,* 33. doi:10.3389/fnhum.2014.00033

Corpus Callosum: Fox, K. C., Nijeboer, S., Dixon, M. L., Floman, J. L., Ellamil, M., Rumak, S. P., Sedlmeier, P., & Christoff, K. (2014). Is meditation associated with altered brain structure? A systematic review and meta-analysis of morphometric neuroimaging in meditation practitioners. *Neuroscience and Biobehavioral Reviews, 43,* 48–73. doi:10.1016/j.neubiorev.2014.03.016

Gyrification: Luders, E., Kurth, F., Mayer, E. A., Toga, A. W., Narr, K. L., & Gaser, C. (2012). The unique brain anatomy of meditation practitioners: Alterations in cortical gyrification. *Frontiers in Human Neuroscience, 6,* 34.

Meditation increases the volume of gray matter: Last, N., Tufts, E., & Auger, L. E. (2017). The effects of meditation on grey matter atrophy and neurodegeneration: A systematic review. *Journal of Alzheimer's Disease, 56*(1), 275–86. doi:10.3233/JAD-160899

Increased cortical thickness associated with three types of meditation studied by the Planck Institute: Goleman, D., & Davidson, R. J. (2018). *Altered traits: Science reveals how meditation changes your mind, brain, and body.* New York, NY: Avery.

Nucleus accumbens shrinks: Goleman, D., & Davidson, R. J. (2018). *Altered traits: Science reveals how meditation changes your mind, brain, and body.* New York, NY: Avery.

Selfing Control Network References

Stronger connection in meditators between the dorsolateral prefrontal cortex and the posterior cingulate cortex: Brewer, J. A., Worhunsky, P. D., Gray, J. R., Tang, Y. Y., Weber, J., & Kober, H. (2011). Meditation experience is associated with differences in default mode network activity and connectivity. *Proceedings of the National Academy of Sciences, 108*(50), 20254-20259.

Brain regions that grow:

Prefrontal cortex: Lazar, S. W., Kerr, C. E., Wasserman, R. H., Gray, J. R., Greve, D. N., Treadway, M. T., . . . Fischl, B. (2005). Meditation experience is associated with increased cortical thickness. *Neuroreport, 16*(17), 1893–1897.

Gotink, R. A., Meijboom, R., Vernooij, M. W., Smits, M., & Hunink, M. M. (2016). 8-week mindfulness based stress reduction induces brain changes similar to traditional long-term meditation practice–a systematic review. *Brain and Cognition, 108,* 32–41. doi:10.1016/j.bandc.2016.07.001

Ventromedial prefrontal cortex: Chau, B. K., Keuper, K., Lo, M., So, K. F., Chan, C. C., & Lee, T. M. (2018). Meditation-induced neuroplastic changes of the prefrontal network are associated with reduced valence perception in older people. *Brain and Neuroscience Advances, 2.* doi:10.1177/2398212818771822

Inferior frontal sulcus and inferior frontal junction: Chau, B. K., Keuper, K., Lo, M., So, K. F., Chan, C. C., & Lee, T. M. (2018). Meditation-induced neuroplastic changes of the prefrontal network are associated with reduced valence perception in older people. *Brain and Neuroscience Advances, 2.* doi:10.1177/2398212818771822

Anterior and mid cingulate: Fox, K. C., Nijeboer, S., Dixon, M. L., Floman, J. L., Ellamil, M., Rumak, S. P., Sedlmeier, P. &, Christoff, K. (2014). Is meditation associated with altered brain structure? A systematic review and meta-analysis of morphometric neuroimaging in meditation practitioners. *Neuroscience and Biobehavioral Reviews, 43,* 48–73. doi:10.1016/j.neubiorev.2014.03.016

Orbitofrontal cortex: Fox et al., 2014. Luders, E., Toga, A. W., Lepore, N., & Gaser, C. (2009). The underlying anatomical correlates of long-term meditation: Larger hippocampal and frontal volumes of gray matter. *Neuroimage, 45*(3), 672–678.

Increased capacity for regulating the DMN just 72 hours after starting meditation practice: Creswell, J. D., Taren, A. A., Lindsay, E. K., Greco, C. M., Gianaros, P. J., Fairgrieve, A., . . . Ferris, J. L. (2016). Alterations in resting-state functional connectivity link mindfulness meditation with reduced interleukin-6: a randomized controlled trial. *Biological Psychiatry, 80*(1), 53–61.

DMN is better controlled even in non-meditating states: Jang, J. H., Jung, W. H., Kang, D. H., Byun, M. S., Kwon, S. J., Choi, C. H., & Kwon, J. S. (2011). Increased default mode network connectivity associated with meditation. *Neuroscience Letters, 487*(3), 358–362.

Empathy Network References

Insula enlarges: Fox, K. C., Nijeboer, S., Dixon, M. L., Floman, J. L., Ellamil, M., Rumak, S. P., Sedlmeier, P., & Christoff, K. (2014). Is meditation associated with altered brain structure? A systematic review and meta-analysis of morphometric neuroimaging in meditation practitioners. *Neuroscience and Biobehavioral Reviews, 43,* 48–73. doi:10.1016/j.neubiorev.2014.03.016

Lazar, S. W., Kerr, C. E., Wasserman, R. H., Gray, J. R., Greve, D. N., Treadway, M. T., . . . Fischl, B. (2005). Meditation experience is associated with increased cortical thickness. *Neuroreport, 16*(17), 1893–1897.

Gotink, R. A., Meijboom, R., Vernooij, M. W., Smits, M., & Hunink, M. M. (2016). 8-week mindfulness based stress reduction induces brain changes similar to traditional long-term meditation practice–a systematic review. *Brain and Cognition, 108,* 32–41. doi:10.1016/j.bandc.2016.07.001

Luders, E., Kurth, F., Mayer, E. A., Toga, A. W., Narr, K. L., & Gaser, C. (2012). The unique brain anatomy of meditation practitioners: Alterations in cortical gyrification. *Frontiers in Human Neuroscience, 6,* 34.

Activation of the TPJ: Goleman & Davidson, 2018.

ACC lights up: Lockwood, P. L., Apps, M. A., Valton, V., Viding, E., & Roiser, J. P. (2016). Neurocomputational mechanisms of prosocial learning and links to empathy. *Proceedings of the National Academy of Sciences, 113*(35), 9763-9768.

Premotor cortex lights up and nucleus accumbens shrinks: Goleman, D., & Davidson, R. J. (2018). *Altered traits: Science reveals how meditation changes your mind, brain, and body.* New York, NY: Avery.

Chapter 7

Veterans Stress Project: Church, D. (2017). *Psychological trauma: Healing its roots in brain, body, and memory* (2nd ed.). Santa Rosa, CA: Energy Psychology Press.

In the wake of traumatic events, some people become more resilient: Tedeschi, R. G., & Calhoun, L. G. (2004). Posttraumatic growth: Conceptual foundations and empirical evidence. *Psychological Inquiry, 15*(1), 1–18.

One third of veterans develop PTSD, but two thirds do not: Tanielian, T. L., & Jaycox, L. H. (Eds.). (2008). *Invisible wounds of war: Psychological and cognitive injuries, their consequences, and services to assist recovery.* Santa Monica, CA: Rand.

Research reveals a correlation between negative childhood events and the development of adult PTSD: Ozer, E. J., Best, S. R., Lipsey, T. L., & Weiss, D. S. (2008). Predictors of posttraumatic stress disorder and symptoms in adults: A meta-analysis. *Psychological Trauma: Theory, Research, Practice, and Policy, 5*(1), 3–36.

Study of 218 veterans and their spouses attending a weeklong retreat: Church, D., & Brooks, A. J. (2014). CAM and energy psychology techniques remediate PTSD symptoms in veterans and spouses. *Explore: The Journal of Science and Healing, 10*(1), 24–33. doi:10.1016/j.explore.2013.10.006

Viktor Frankl: Burton, N. (2012, May 24). Man's search for meaning: Meaning as a cure for depression and other ills. *Psychology Today.* Retrieved Jan 2, 2020 from https://www.psychologytoday.com/us/blog/hide-and-seek/201205/mans-search-meaning

Frankl, V. (1959). *Man's search for meaning.* Boston, MA: Beacon Press.

About 75% of Americans will experience a traumatic event: Joseph, S. (2011). *What doesn't kill us: The new psychology of posttraumatic growth.* New York, NY: Basic Books.

Women are more likely to be victims of domestic violence: Van der Kolk, B. A. (2014). *The body keeps the score: Brain, mind, and body in the healing of trauma.* New York, NY: Viking.

One in 10 boys is molested and 1 in 5 girls: Gorey, K. M., & Leslie, D. R. (1997). The prevalence of child sexual abuse: Integrative review adjustment for potential response and measurement biases. *Child Abuse and Neglect, 21*(4), 391–398.

60% of teenagers witness or experience victimization: U.S. Department of Health and Human Services. (2012). *Child maltreatment 2011.* Washington, DC: Administration for Children and Families, Administration on Children, Youth and Families, Children's Bureau. Retrieved from http://www.acf.hhs.gov

More Americans died at the hands of family members: Van der Kolk, B. A. (2014). *The body keeps the score: Brain, mind, and body in the healing of trauma.* New York, NY: Viking.

Biological embedding: Shonkoff, J. P., Boyce, W. T., & McEwen, B. S. (2009). Neuroscience, molecular biology, and the childhood roots of health disparities: Building a new framework for health promotion and disease prevention. *JAMA, 301*(21), 2252–2259.

RCT of veterans with PTSD with microRNAs bound to DNA: Yount, G., Church, D., Rachlin, K., Blickheuser, K., & Cardonna, I. (2019). Do noncoding RNAs mediate the efficacy of Energy Psychology? *Global Advances in Health and Medicine, 8,* 2164956119832500.

VMPFC and DLPFC control the amygdala: Goleman, D., & Davidson, R. J. (2018). *Altered traits: Science reveals how meditation changes your mind, brain, and body.* New York, NY: Avery.

The more hours of practice, the stronger the effect: Goleman, D., & Davidson, R. J. (2018). *Altered traits: Science reveals how meditation changes your mind, brain, and body.* New York, NY: Avery Goleman & Davidson, *Altered traits.*

Amygdala reductions of as much as 50%: Goleman, D., & Davidson, R. J. (2018). *Altered traits: Science reveals how meditation changes your mind, brain, and body.* New York, NY: Avery.

Linda Graham's exercises to build a resilient brain: Graham, L. (2019). *Resilience: Powerful practices for bouncing back from disappointment, difficulty, and even disaster.* Novato, CA: New World Library.

Bruce McEwen regards resilience as an internally generated state: McEwen, B. S. (2016). In pursuit of resilience: stress, epigenetics, and brain plasticity. *Annals of the New York Academy of Sciences, 1373*(1), 56–64.

Resilience is more: Hanson, R., & Hanson, F. (2018). *Resilient: How to grow an unshakable core of calm, strength, and happiness.* New York, NY: Harmony Books.

Linda Graham advocates "coming to see ourselves as people who can be resilient": Graham, L. (2019). *Resilience: Powerful practices for bouncing back from disappointment, difficulty, and even disaster.* Novato, CA: New World Library.

A 15-year study of professional failure: Wang, Y., Jones, B. F., & Wang, D. (2019). Early-career setback and future career impact. *Nature Communications, 10*(1) DOI: 10.1038/s41467-019-12189-3

Between 35% and 65% of people who experience a disaster return to their normal routine: Jaffe, E. (2012, July/August). *A glimpse inside the brains of trauma survivors.* Association for Psychological Science. Retrieved from https://www.psychologicalscience.org/observer/the-psychology-of-resilience

Boston University School of Medicine optimism study: Lee, L. O., James, P., Zevon, E. S., Kim, E. S., Trudel-Fitzgerald, C., Spiro, A., . . . Kubzansky, L. D. (2019). Optimism is associated with exceptional longevity in 2 epidemiologic cohorts of men and women. *Proceedings of the National Academy of Sciences, 116*(37), 18357–18362.

Studies show that meditators improve on a wide array of biomarkers: Pelletier, K. R. (2018). *Change your genes, change your life: Creating optimal health with the new science of epigenetics.* San Francisco, CA: Red Wheel/Weiser.

Chapter 8

Classical violinists, elite athletes, and race car drivers: Church, D. (2013). *The genie in your genes.* Fulton, CA: Energy Psychology.

The definition of selective attention: Raz, A. (2004). Anatomy of attentional networks. *Anatomical Record Part B: The New Anatomist, 281*(1), 21–36.

Training attention amidst the distractions and annoyances of everyday life: Baumeister, R. F., DeWall, C. N., Vohs, K. D., & Alquist, J. L. (2010). Does emotion cause behavior (apart from making people do stupid, destructive things)? In C. R. Agnew, D. E. Carlston, W. G. Graziano, & J. R. Kelly (Eds.), *Then a miracle occurs: Focusing on behavior in social psychological theory and research* (pp. 119–136). New York, NY: Oxford University Press.

Identical twins die more than 10 years apart: Church, D. (2013). *The genie in your genes.* Fulton, CA: Energy Psychology.

The brains of adepts age more slowly: Goleman, D., & Davidson, R. J. (2018). *Altered traits: Science reveals how meditation changes your mind, brain, and body.* New York, NY: Avery.

Global well-being is rapidly improving: Pinker, S. (2019). *Enlightenment now: The case for reason, science, humanism, and progress.* New York, NY: Penguin.

Deforestation of the Amazon rainforest dropped by 80% in the past 2 decades: Boucher, D. (2014). How Brazil has dramatically reduced tropical deforestation. *Solutions Journal* 5(2), 66-75.

The American poverty rate from 1960 through today: Economist (2018; March 1). Poverty in America: A never-ending war. Retrieved Nov 26, 2019 from: https://www.economist.com/democracy-in-america/2018/03/01/poverty-in-america

A scientific review of the role that compassion plays in evolution: Goetz, J. L., Keltner, D., & Simon-Thomas, E. (2010). Compassion: an evolutionary analysis and empirical review. *Psychological Bulletin, 136*(3), 351.

Darwin and the evolution of emotion: DiSalvo, D. (2009, Feb 26). Forget survival of the fittest: It is kindness that counts. *Scientific American* blog. Retrieved Jan 4, 2020 from https://www.scientificamerican.com/article/kindness-emotions-psychology

The number of meditation practitioners tripled between 2012 and 2017: Clarke, T. C., Barnes, P. M., Black, L. I., Stussman, B. J., & Nahin, R. L. (2018). *Use of yoga, meditation, and chiropractors among U.S. adults aged 18 and older.* NCHS Data Brief, no 325. Hyattsville, MD: National Center for Health Statistics.

European Values Study. European Values Study (n.d.). Religion. Retrieved Jan 9, 2020 from Retrieved from https://europeanvaluesstudy.eu/about-evs/research-topics/religion

Reagan's second inaugural address: Lillian Goldman Law Library (2008). Second inaugural address of Ronald Reagan. Retrieved Jan 2, 2020 from https://avalon.law.yale.edu/20th_century/reagan2.asp

David Houle and the "Finite Earth Economy." Houle, D. (2019). Moving to a Finite Earth Economy. Presentation at Transformational Leadership Council, San Diego, CA. July 25, 2019.

Sea level rise predictions: Meyer, R. (2019, January 4). A Terrifying Sea-Level Prediction Now Looks Far Less Likely: But experts warn that our overall picture of sea-level rise looks far scarier today than it did even five years ago. *The Atlantic.* Retrieved Jan 12, 2020 from https://www.theatlantic.com/science/archive/2019/01/sea-level-rise-may-not-become-catastrophic-until-after-2100/579478

Sea level rise and Bangladesh: Glennon, R. (2017, April 21). The Unfolding Tragedy of Climate Change in Bangladesh. *Scientific American.* Retrieved Jan 2, 2020 from https://blogs.scientificamerican.com/guest-blog/the-unfolding-tragedy-of-climate-change-in-bangladesh

Daniel Schmachtenberger and weaponized AI: Schmachtenberger, D. (2019). Winning at the wrong game. Presentation at Transformational Leadership Council, San Diego, CA. July 27, 2019.

Over 1 million species of animals and plants are at risk of extinction: CBS News (2019, May 6). One million species of plants and animals at risk of extinction, UN report warns. Retrieved Jan 4, 2020 from https://www.cbsnews.com/news/report-1-million-animals-plant-species-face-extinction-due-climate-change-human-activity-population

The $6 trillion cost of the Middle Eastern war: Crawford, N. (2016). *US budgetary costs of wars through 2016: $4.79 trillion and counting.* Watson Institute, Brown University. Retrieved Jan 4, 2020 from from http://watson. brown . edu/costsofwar/files/cow/imce/papers/2016/Costs

Global debt is over $200 trillion, three times the size of the global economy: Durden, T. (2019). *Global Debt Hits $246 Trillion, 320% Of GDP, As Developing Debt Hits All Time high.* ZeroHedge. Retrieved November 21, 2019, from https://www.zerohedge.com/news/2019-07-15/global-debt-hits-246-trillion-320-gdp-developing-debt-hit-all-time-high

Over 2 billion human beings don't have access to clean drinking water: World Health Organization (2019: June 18). 1 in 3 people globally do not have access to safe drinking water. Retrieved Jan 2, 2020 from https://www.who.int/news-room/detail/18-06-2019-1-in-3-people-globally-do-not-have-access-to-safe-drinking-water-%E2%80%93-unicef-who

In the past 50 years, 29% of the birds in North America have disappeared: Rosenberg, K. V., Dokter, A. M., Blancher, P. J., Sauer, J. R., Smith, A. C., Smith, P. A., . . . Marra, P. P. (2019). Decline of the North American avifauna. *Science, 366*(6461), 120-124.

Millennials and global problems: Jackson, A. (2017, Aug 29). These are the world's 10 most serious problems, according to millennials. *Inc* magazine blog. Retrieved Jan 11, 2020 from https://www.inc.com/business-insider/worlds-top-10-problems-according-millennials-world-economic-forum-global-shapers-survey-2017.html

Emotions spread through communities like the flu: Hatfield, E., Cacioppo, J. T., & Rapson, R. L. (1994). *Emotional contagion.* New York, NY: Cambridge University Press.

Chapman, R., & Sisodia, R. (2015). *Everybody matters: The extraordinary power of caring for your people like family.* New York, NY: Penguin.

Fowler, J. H., & Christakis, N. A. (2008). Dynamic spread of happiness in a large social network: Longitudinal analysis over 20 years in the Framingham Heart Study. *British Medical Journal, 337,* a2338.

Experimenters were able to induce emotional contagion in Facebook users: Kramer, A. D., Guillory, J. E., & Hancock, J. T. (2014). Experimental evidence of massive-scale emotional contagion through social networks. *Proceedings of the National Academy of Sciences, 111*(24), 8788–8790.

Volunteers going into heart coherence were able to induce coherence in test subjects: Morris, S. M. (2010). Achieving collective coherence: Group effects on heart rate variability coherence and heart rhythm synchronization. *Alternative Therapies in Health and Medicine, 16*(4), 62–72.

Field effects research: HeartMath Institute. (n.d.). *Global coherence research: The science of interconnectivity.* Retrieved Jan 2, 2020 from https://www.heartmath.org/research/global-coherence

We're all like little cells in the bigger Earth brain: McCraty, R., Atkinson, M., & Bradley, R. T. (2004). Electrophysiological evidence of intuition: Part 1. The surprising role of the heart. *The Journal of Alternative & Complementary Medicine, 10*(1), 133-143.

The Caltech experiment showing humans sense the planet's fields: Wang, C. X., Hilburn, I. A., Wu, D. A., Mizuhara, Y., Cousté, C. P., Abrahams, J. N., . . . Kirschvink, J. L. (2019). Transduction of the geomagnetic field as evidenced from alpha-band activity in the human brain. *eneuro, 6*(2), 4-19.

Barack Obama speech at Gates Foundation conference: Weller, C. (2017, September 20). Barack Obama has a one-question test that proves how good the world is today. *Business Insider.* Retrieved Jan 2, 2020 from https://www.businessinsider.com/president-barack-obama-speech-goalkeepers-2017-9

INDEX

NOTE: Page references in *italics* refer to illustrations and photos.

C

D

E

F

G

H

I

J

K

L

M

N

S

W

X

Y

Z

IMAGE CREDITS

1.8. By Planet Labs, Inc - Own work, CC BY-SA 4.0, https://commons.wikimedia.org/wiki/File:Santa_Rosa_4Mar2018_SkySat.jpg

Rumi poem "The Guest House" from the book *The Essential Rumi*, translations by Coleman Barks, used with permission of HarperCollins and Coleman Barks.

2.2. © Can Stock Photo / edharcanstock

2.5. By Andreashorn - Own work, CC BY-SA 4.0, https://upload.wikimedia.org/wikipedia/commons/5/57/Default_Mode_Network_Connectivity.png

2.6. By Hintha - Own work, CC BY-SA 3.0, https://commons.wikimedia.org/w/index.php?curid=11359793

2.10. © Can Stock Photo / tomwang

4.2. By Dennis Jarvis - Own work, CC BY-SA 2.0, https://commons.wikimedia.org/wiki/File:Jokhang_dharma_wheel-5447.jpg

4.5. By Morn - Own work, CC BY-SA 3.0, https://commons.wikimedia.org/w/index.php?curid=18949373

4.22. By Ailia Jameel - Own work, CC BY-SA 4.0, https://commons.wikimedia.org/wiki/File:Classroom_lessons.jpg

4.24. By Sabh Benziadi - Own work, CC BY-SA 4.0, https://commons.wikimedia.org/wiki/File:Sufi_palazzo_dei_Normanni_Palermo_con_Sabah_Benziadi.jpg

4.26. © Istockphoto / Fatcamera

4.30. Photo © Alex Ze'evi Christian; website: alexzphotography.com. Reprinted with permission from Guy Spier and Alex Ze'evi Christian.

5.1. © Istockphoto / Nerthuz

5.6 By Dannybalanta - Own work, CC BY-SA 3.0, https://commons.wikimedia.org/w/index.php?curid=31432711

5.20. © Istockphoto / nyshooter

5.22. © Istockphoto / AntonioGuillem

6.6. © Istockphoto / CasarsaGuru

6.7. © Istockphoto / ArtMarie

6.9. Used with permission of Dr. Kinichi Nakashima and Dr. Masakazu Namihira.

7.1. © Istockphoto / vm

7.3. By Christoph Bock, Max Planck Institute for Informatics - Own work, CC BY-SA 3.0, https://commons.wikimedia.org/w/index.php?curid=17066877

7.5. Soldier: © Istockphoto / Highwaystarz-Photography

MicroRNA: By Ppgardne at en.wikipedia, CC BY-SA 3.0, https://commons.wikimedia.org/w/index.php?curid=16639717

8.1. © Istockphoto / fcafotodigital

8.2. © Istockphoto / BorupFoto

8.3. © Istockphoto / middelveld

8.4. By Gazebo - Own work, CC BY-SA 3.0, https://commons.wikimedia.org/w/index.php?curid=49467838

8.5. CNX OpenStax - CC BY 4.0, https://commons.wikimedia.org/w/index.php?curid=49931564

8.6. Max Roser - https://ourworldindata.org/life-expectancy, CC BY 4.0, https://commons.wikimedia.org/w/index.php?curid=83546093

8.7. Max Roser - https://ourworldindata.org/life-expectancy, CC BY 4.0, https://commons.wikimedia.org/w/index.php?curid=83546093

8.8. By Alfie↑↓© - Own work, CC BY-SA 3.0, https://commons.wikimedia.org/w/index.php?curid=11481813

ACKNOWLEDGMENTS

You can't delegate a haircut. While the CEO of a $10 billion corporation or the US president can delegate almost any task, neither can tell someone, "Go get a haircut for me." You have to be there in person.

Meditation is like that. You can't hire a gifted surrogate to meditate for you. And no matter how high the heights a saint has reached, you can't port their state into your brain. Meditation is one of the great levelers. You have to achieve those states all by yourself. My intention is that *Bliss Brain* has helped you get there.

Science is the opposite. Each scientific discovery builds on those that came before. We "stand on the shoulders of giants." We gather the threads of brilliance and take them to the next level. The next generation starts building on the top of the highrise, not on the ground floor.

I have been indebted to the thousands of scientists whose work has provided the foundation for this book. Hundreds of these are listed in the endnotes. In the references to their works, you'll find thousands more. These are the giants on whose shoulders *Bliss Brain* is built.

There are also several absolutely brilliant books that have inspired sections of this one. These are the thinkers who constructed the floors on which *Bliss Brain* is built. I'm fortunate to have met or interviewed many of them.

Science writing of the sort we do is a demanding endeavor. You read the research of others, and seek to explain it to a general audience. This means studying papers with titles like "DNA methyltransferase 1 is indispensable for development of the hippocampal dentate gyrus."

You have to do more than understand the highly technical jargon they contain. You're required to determine if they're relevant, then explain them to readers in a way that makes sense. You also have to weave your sources into a meaningful scientific story that shows how all the parts fit together.

There are three outstanding books that do this magnificently. I drew on them heavily when writing *Bliss Brain*. They are:

Stealing Fire: How Silicon Valley, the Navy SEALs, and Maverick Scientists Are Revolutionizing the Way We Live and Work by Steven Kotler and Jamie Wheal.

Altered Traits: Science Reveals How Meditation Changes Your Mind, Brain, and Body by Daniel Goleman and Richard Davidson.

How Enlightenment Changes Your Brain: The New Science of Transformation by Andrew Newberg and Mark Robert Waldman.

These three books are audacious in their scope and profundity. Not only are they packed with astonishing science, they're beautifully written. They are books to savor. I recommend you read them carefully, delighting in each page. Read them more than once. Other key ideas in *Bliss Brain* came from these excellent works:

Behave: The Biology of Humans at Our Best and Worst by Robert M. Sapolsky.

Aware: The Science and Practice of Presence—The Groundbreaking Meditation Practice by Daniel Siegel.

Resilience: Powerful Practices for Bouncing Back from Disappointment, Difficulty, and Even Disaster by Linda Graham.

Resilient: How to Grow an Unshakable Core of Calm, Strength, and Happiness by Rick Hanson and Forrest Hanson.

Enlightenment Now: The Case for Reason, Science, Humanism, and Progress by Steven Pinker.

Several friends read all or part of the manuscript and gave me feedback. Such informed opinion is like gold. Yet you're asking some of the world's busiest people to take days of their time and provide you with detailed feedback. That's a big ask, and I'm very grateful to those who helped mature the text and detect my errors of understanding or fact.

David Feinstein, PhD, co-author of *Energy Medicine* and several other award-winning books, read the whole book and gave me line-by-line feedback. Every single chapter is better and clearer as a result of his comments.

Bob Hoss is the co-editor of *Dreams That Change Our Lives*, a longtime friend, and an NIIH board member. He gave me pointed feedback on the brain function and structure chapters, and was especially diligent in pointing out the useful functions of the DMN, preventing me from turning it into an unalloyed villain.

John Dupuy of iAwake Technologies read the chapters on brain function and structure to make sure I made no egregious errors. Gary Groesbeck is an EEG expert with a vast knowledge of brain function, and commented on several chapters. So did Mind Mirror expert Judith Pennington. Gary, Judith, and Gary's wife, Donna Bach, have run EEG experiments at several of my workshops, providing empirical validation of the peak states participants experience there. Their warm and caring presence has touched me deeply, as well as those they serve.

I'm grateful for the mentorship of Lisa Nichols. Her speaking workshop improved my storytelling. She's a member of Jack Canfield's Transformational Leadership Council or TLC, as am I. I'm grateful to the many TLC members who read early drafts of the book and wrote endorsements.

It seems surreal that in the year after the fire, and during the health and financial crashes that followed, I wrote a book called *Bliss Brain*. I still can't quite believe it. Without my dear friend and longtime editor Stephanie Marohn, the book would never have happened. She wrote the first drafts of several chapters when I had not an hour in a month to devote to writing.

Kira Theine has been transcribing audios for me for several years. Chapter 1 was dictated into my iPhone while driving to a retreat, and sent to Kira for transcription. So have many of my blog posts and articles.

I appreciate my editor at Hay House, Anne Barthel, as well as publisher Patty Gift and the many people on the Hay House production team. CEO Reid Tracy is both a trusted friend and a brilliant professional mentor.

I'm also grateful to Jennifer Ellis, my promotion manager, and Christy Granzotto, my social media manager, for their longtime support and eternal positivity. Cathy Veloskey and Lindsay McGinty of the Hay House promotional team have been enthusiastic and imaginative cheerleaders for getting the message of *Bliss Brain* out into the world.

I have a wonderful team of people I work with at Energy Psychology Group / EFT Universe. Heather Montgomery managed them for 3 years, freeing me up to write both *Mind to Matter* and *Bliss Brain*. Thank you, Seth, Kendra, Marion, and Jackie, for handling so many details so I didn't have to.

I'm indebted to Dave Asprey for being a funny and wise friend, for getting together with me a number of times, and for writing the foreword. Dave is the uncrowned king of "biohacking." That's the science of tuning your mind and body for optimal longevity and performance. Dave was my first choice to write the foreword because I wanted to make it clear how profound a biohack meditation can be.

When I meditate each morning, I tune in to the huge community of people who practice each day. I feel one in consciousness with them and the universe. I am grateful for the sense of connection that meditators share, and out of which the inspiration for *Bliss Brain* sprang. We're motivated by the vision of every being abandoning their suffering and joining us in bliss.

ABOUT THE AUTHOR

Dawson Church is an award-winning author whose bestselling book *The Genie in Your Genes* (www.YourGeniusGene.com) has been hailed by reviewers as a breakthrough in our understanding of the link between emotions and genetics. His follow-up title, *Mind to Matter* (www.MindToMatter.com), reviews the science of peak mental states.

He founded the National Institute for Integrative Healthcare (www.NIIH.org) to study and implement promising evidence-based psychological and medical techniques. In his undergraduate and graduate work at Baylor University, he became the first student to successfully graduate from the academically rigorous University Scholars program in 1979. He earned a doctorate at Holos University under the mentorship of neurosurgeon Norman Shealy, MD, PhD, founder of the American Holistic Medical Association.

After an early career in book publishing as editor and then president of Aslan Publishing (www.aslanpublishing.com), Church went on to receive a postgraduate PhD in Natural Medicine as well as clinical certification in Energy Psychology. Church's ground-breaking research has been published in prestigious scientific journals. He is the editor of *Energy Psychology: Theory, Research & Treatment,* a peer-reviewed professional journal.

He shares how to apply the breakthroughs of energy psychology to health and athletic performance through EFT Universe (www.EFTUniverse.com). EFT Universe was the first organization to have its courses accredited for CME (continuing medical education) for all the major professions, including doctors (AMA), psychologists (APA), and nurses (ANCC). He has trained thousands of practitioners in energy psychology techniques and offers the premier certification program in the field (EnergyPsychologyCertification.com). **www.dawsonchurch.com**

Hay House Titles of Related Interest

YOU CAN HEAL YOUR LIFE, the movie, starring Louise Hay & Friends (available as a 1-DVD program, an expanded 2-DVD set, and an online streaming video)
Learn more at www.hayhouse.com/louise-movie

THE SHIFT, the movie,
starring Dr. Wayne W. Dyer
(available as a 1-DVD program, an expanded 2-DVD set, and an online streaming video)
Learn more at www.hayhouse.com/the-shift-movie

***

BECOMING SUPERNATURAL: How Common People Are Doing the Uncommon, by Dr. Joe Dispenza

THE SCIENCE BEHIND TAPPING: A Proven Stress Management Technique for the Mind and Body, by Peta Stapleton, PhD

THE SCIENCE OF SELF-EMPOWERMENT: Awakening the New Human Story, by Gregg Braden

All of the above are available at www.hayhouse.co.uk

***

CONNECT WITH
HAY HOUSE
ONLINE

'The gateways to wisdom and knowledge are always open.'

Louise Hay